NUTRITION AND DIET RESEARCH PROGRESS

MICRONUTRIENTS AND THEIR ROLE IN HEALTH AND DISEASE

NUTRITION AND DIET RESEARCH PROGRESS

Additional books and e-books in this series can be found on Nova's website under the Series tab.

NUTRITION AND DIET RESEARCH PROGRESS

MICRONUTRIENTS AND THEIR ROLE IN HEALTH AND DISEASE

HORACE A. HOWARD
EDITOR

Copyright © 2021 by Nova Science Publishers, Inc.

All rights reserved. No part of this book may be reproduced, stored in a retrieval system or transmitted in any form or by any means: electronic, electrostatic, magnetic, tape, mechanical photocopying, recording or otherwise without the written permission of the Publisher.

We have partnered with Copyright Clearance Center to make it easy for you to obtain permissions to reuse content from this publication. Simply navigate to this publication's page on Nova's website and locate the "Get Permission" button below the title description. This button is linked directly to the title's permission page on copyright.com. Alternatively, you can visit copyright.com and search by title, ISBN, or ISSN.

For further questions about using the service on copyright.com, please contact:
Copyright Clearance Center
Phone: +1-(978) 750-8400   Fax: +1-(978) 750-4470   E-mail: info@copyright.com

NOTICE TO THE READER

The Publisher has taken reasonable care in the preparation of this book, but makes no expressed or implied warranty of any kind and assumes no responsibility for any errors or omissions. No liability is assumed for incidental or consequential damages in connection with or arising out of information contained in this book. The Publisher shall not be liable for any special, consequential, or exemplary damages resulting, in whole or in part, from the readers' use of, or reliance upon, this material. Any parts of this book based on government reports are so indicated and copyright is claimed for those parts to the extent applicable to compilations of such works.

Independent verification should be sought for any data, advice or recommendations contained in this book. In addition, no responsibility is assumed by the Publisher for any injury and/or damage to persons or property arising from any methods, products, instructions, ideas or otherwise contained in this publication.

This publication is designed to provide accurate and authoritative information with regard to the subject matter covered herein. It is sold with the clear understanding that the Publisher is not engaged in rendering legal or any other professional services. If legal or any other expert assistance is required, the services of a competent person should be sought. FROM A DECLARATION OF PARTICIPANTS JOINTLY ADOPTED BY A COMMITTEE OF THE AMERICAN BAR ASSOCIATION AND A COMMITTEE OF PUBLISHERS.

Additional color graphics may be available in the e-book version of this book.

Library of Congress Cataloging-in-Publication Data

ISBN: 978-1-53619-843-0

Published by Nova Science Publishers, Inc. † New York

CONTENTS

Preface		vii
Chapter 1	Micronutrients in Chronic Kidney Disease *Oguzhan Sıtkı Dizdar and Alparslan Ersoy*	1
Chapter 2	Phytochemical Deflection of Harmful Inflammatory Events toward More Effective Immune Activity *Stephen C. Bondy*	51
Chapter 3	Micronutrients Needs in Critically Ill Patients of Pediatric Care Units *Nazanin Zibanejad*	81
Index		103

PREFACE

Micronutrients, which include vitamins and minerals, are an essential component of any healthy diet. Though the body only requires small amounts of micronutrients, they nonetheless play a critical role in health and disease. The first chapter of this book includes a review of the relationships between micronutrient levels and clinical outcomes in patients with chronic kidney disease and the effects of different types of renal replacement therapy on micronutrient levels. The second chapter focuses on the use of micronutrient phytochemicals for optimizing health in normal subjects to assure they have the best possible immune response to an adverse disease event. Finally, the third chapter summarizes the data on the possible benefits and harms of vitamin supplements and essential minerals in pediatric ICU patients.

Chapter 1 - Chronic Kidney Disease (CKD) is a leading public health problem that has received increased research attention because of its significantly increased prevalence. These patients are at risk of malnutrition due to several factors, including chronic comorbidities, poor nutritional intake, and uremia. It is known that malnutrition increases morbidity and mortality, possibly due to specific micronutrient deficiencies. In patients treated with renal replacement therapy (RRT), unrecognized losses of nutrients in dialysate or urine, poor food intake, intestinal dysfunction, inflammation, and abnormal metabolism may also contribute to micronutrient deficiencies. Some micronutrients are negatively associated with CKD stages. On the other hand, some other micronutrients' plasma concentration increases, paralleling the rise in creatinine concentration. CKD is characterized by inflammation due to oxidative stress. Many micronutrients are components of metalloenzymes and participate in reactive oxygen metabolism, free radical scavenging, and hormone activities. Therefore, the micronutrient levels are significant in ensuring optimal management among CKD patients with or without RRT. Different types of RRT may not have a similar effect on micronutrient levels. It is unclear whether all alterations of micronutrient status contribute to the poor outcomes of CKD patients. Also, it is unclear whether routine micronutrient intake (vitamins and trace elements) affects outcomes in adults with CKD 1-5D, non-dialysis, and transplant. As a result, there are many controversial results in the literature, and more information about the micronutrient status of CKD patients receiving RRT is needed. Here, the authors aimed to review the relationships between micronutrient levels and clinical outcomes in CKD patients and the effects of different types of RRT on micronutrient levels.

Chapter 2 - A large number of disorders are associated with two changes in immune reactivity. Firstly, a flourishing disease is often accompanied by depression of effective immune surveillance. This shortcoming allows both invasive microorganisms to proliferate and also leads to survival of abnormal cell types, some of which may proliferate and become transformed to malignant variants. A second feature common to several chronic disease states, is the development of ineffective immune reactions, largely typified by elevation of non-selective generalized inflammatory events. Both of these characteristics adversely affect control and limitation of disease progression. These deficits comprising ever evolving levels of inflammation and a poverty of beneficial immune responses, are also found in normal aging and consequently are especially marked in age-related ailments. The basis for the use of micronutrient phytochemicals is to optimize health in normal subjects and thus pre-emptively assure that the response to an adverse disease event will be the best possible. The maintenance of a high level of fitness in advance of untoward circumstances, is important in delaying and limiting the consequences of both disease states and changes accompanying senescence. This chapter emphasizes the utility of a series of micronutrient phytochemicals that are able to redirect immune responses toward a more specific goal, while diminishing the propensity of inflammatory mechanisms to be over-reactive and inappropriately directed. In general, phytonutrients tend to be less directed toward a single metabolic site than are pharmacological agents, and tend to impact a series of targets. While they are not essential like vitamins and essential minerals, they are likely to have substantial benefits regarding overall human health and longevity as a result of their disease-preventing properties. This breadth of action and generally low toxicity is in contrast to pharmaceuticals which carry a higher risk of adverse side effects. This makes phytochemicals eminently suitable for extended usage over a prolonged period.

Chapter 3 - Micronutrients are essential nutrient needed by the body in small amounts, including minerals and vitamins. Micro minerals include calcium, phosphorus, magnesium, sodium, potassium, and trace minerals, including magnesium, iron, zinc and selenium. Vitamins and minerals are vital for the prevention of disease, development and health of the body. Only small amounts of them are needed, but they are not produced in the body and must be received through diet. Deficiency of trace elements such as magnesium zinc phosphate is common in patients admitted to the ICU. Studies have shown that this deficiency is associated with a poor prognosis. Among vitamins for example vitamin C have antioxidant effects improve inflammation and have a beneficial effect. In acute renal failure, meta-analysis has shown that vitamin C intake reduces the ICU stay and the duration of mechanical ventilation. Low level of Vitamin D have been shown to be strongly associated with problems such as infections, acute liver and kidney damage. However, the effect of vitamin D supplementation is unclear and has not improved the prognosis in one meta-analysis. Zinc levels were significantly lower in patients requiring hospitalization in pediatric ICU than in patients with pneumonia admitted to the ward. In this section, the authors will summarize the data of existing articles on the effects of possible benefits and harms of vitamin supplements and essential minerals in pediatric ICU patients and review the results of these studies. Prescribing micronutrients to pediatric ICU patients can have beneficial effects, including reducing hospital and ICU hospital stays, reducing respiratory illness, and improving infections. In this chapter of the book, the authors intend to provide an overview of the topic of micronutrients in critically ill children admitted to the PICU.

In: Micronutrients and their Role...　　ISBN: 978-1-53619-843-0
Editor: Horace A. Howard　　© 2021 Nova Science Publishers, Inc.

Chapter 1

MICRONUTRIENTS IN CHRONIC KIDNEY DISEASE

Oguzhan Sıtkı Dizdar[1] and Alparslan Ersoy[2]

[1]Kayseri City Training and Research Hospital,
Department of Internal Medicine and Clinical Nutrition,
Kayseri, Turkey
[2]Department of Internal Medicine, Division of Nephrology,
Bursa Uludag University Faculty of Medicine, Bursa, Turkey

ABSTRACT

Chronic Kidney Disease (CKD) is a leading public health problem that has received increased research attention because of its significantly increased prevalence. These patients are at risk of malnutrition due to several factors, including chronic comorbidities, poor nutritional intake, and uremia. It is known that malnutrition increases morbidity and mortality, possibly due to specific micronutrient deficiencies. In patients treated with renal replacement therapy (RRT), unrecognized losses of nutrients in dialysate or urine, poor food intake, intestinal dysfunction, inflammation, and abnormal metabolism may also contribute to micronutrient deficiencies. Some micronutrients are negatively associated with CKD stages. On the other hand, some other micronutrients' plasma concentration increases, paralleling the rise in creatinine concentration.

CKD is characterized by inflammation due to oxidative stress. Many micronutrients are components of metalloenzymes and participate in reactive oxygen metabolism, free radical scavenging, and hormone activities. Therefore, the micronutrient levels are significant in ensuring optimal management among CKD patients with or without RRT.

Different types of RRT may not have a similar effect on micronutrient levels. It is unclear whether all alterations of micronutrient status contribute to the poor outcomes of CKD patients. Also, it is unclear whether routine micronutrient intake (vitamins and trace elements) affects outcomes in adults with CKD 1-5D, non-dialysis, and transplant. As a result, there are many controversial results in the literature, and more information about the micronutrient status of CKD patients receiving RRT is needed. Here, we aimed to review the relationships between micronutrient levels and clinical outcomes in CKD patients and the effects of different types of RRT on micronutrient levels.

Keywords: micronutrient, minerals, trace elements, vitamins, renal replacement therapy, hemodialysis, peritoneal dialysis, continuous renal replacement therapy, kidney transplantation, treatment, outcomes

INTRODUCTION

Chronic kidney disease (CKD) is a prevalent chronic condition and has significant health and lifestyle implications for those affected, including increased risk of cardiovascular disease [1] and malnutrition [2]. Malnutrition is associated with disordered micronutrient status and so contributes to morbidity and mortality among CKD patients. Therefore, the micronutrient levels are critical in providing optimal management among non-dialysis CKD patients and patients receiving renal replacement therapy (RRT).

CKD patients may have increased proteinuria, by which there may be an excessive loss of protein-bound micronutrients. Dialysis removes uremic toxins by allowing the equilibration of plasma water and dialysate across a semi-permeable membrane. Theoretically, dialysis patients are at risk for both deficiency and accumulation of trace elements depending upon dietary intake, removal by dialysis, the composition of the source water used for dialysate, and residual renal function. There are some important differences in micronutrient status between hemodialysis patients, continuous ambulatory peritoneal dialysis (CAPD) patients, and kidney transplant patients. Compared to hemodialysis patients, CAPD patients are exposed to much lower dialysate volumes. Therefore it has a different effect on micronutrient levels.

Table 1. Clinical findings that can be seen in the deficiency and/or toxicity of some trace elements [3, 4]

Zinc	Deficiency: growth retardation, delayed sexual maturation, impotence, hypogonadism, oligospermia, hair changes (alopecia), impaired appetite, dysgeusia (impaired taste), night blindness, various skin lesions, decubitus ulcers, delayed wound healing, immune deficiency (impaired cell proliferation, abnormal T-cell function, defective phagocytosis, and abnormal cytokine expression)
	HD: excess risk of infection, anorexia, impaired taste or smell, skin fragility, impotence, impaired cognitive function, peripheral neuropathy
	Toxicity: abdominal pain, diarrhea, nausea, vomiting, copper deficiency
Selenium	Deficiency: skeletal muscle dysfunction, mood disorders, impaired immune function, macrocytosis, whitened nailbeds, sudden death, cardiomyopathy, hypertension, heart failure, coronary artery disease, increased susceptibility to oxidant stress, increased risk of cancer; HD: cardiomyopathy, increased oxidative stress, skeletal myopathy, thyroid dysfunction, hemolysis, dermatosis
	Toxicity: nausea, emesis, diarrhea, hair loss, nail changes, mental status changes, visual loss, peripheral neuropathy, reversible posterior encephalopathy syndrome
Lead	Toxicity: impaired cognitive function, impaired hemoglobin synthesis, hypertension (risk of cardiovascular disease), cerebrovascular disease, renal insufficiency
Arsenic	Toxicity: oxidative injury, inhibition of DNA repair, chromosomal damage (deletion, aneuploidy), peripheral vascular disease
Aluminum	Toxicity: HD: anemia, disabling encephalopathy, neuropathy, severe symptomatic bone disease
Copper	Deficiency: fragile, abnormally-formed hair, depigmentation of the skin, muscle weakness (myeloneuropathy), neurologic abnormalities, edema, hepatosplenomegaly, osteoporosis, neurologic manifestations (ataxia, neuropathy, and cognitive deficits), anemia, thrombocytopenia
	Toxicity: cardiac and renal failure, intravascular hemolysis, hepatic necrosis, encephalopathy
Manganese	Deficiency: poor growth, decreased fertility, ataxia, skeletal deformities, abnormal fat and carbohydrate metabolism
	Toxicity: dyscoordination, loss of balance, confusion, headache, vomiting, hepatic dysfunction

Trace elements are divided into two groups according to the amount of mineral required in adults, ultra-trace (<1 mg/day; copper, iron, manganese, and zinc, etc.) and trace (1 to 100 mg/day; arsenic, chromium, molybdenum, nickel, selenium, and vanadium, etc.). Both deficiency and excess of trace elements are potentially harmful (Table 1) [3].

Deficiency of essential micronutrients and excess of potentially harmful micronutrients are both known to have adverse consequences in the CKD population [5]. However, there are many controversial data about the incidence of abnormal micronutrient status in CKD patients. Micronutrient supplementation is needed to maintain adequate nutrition status for patients with CKD who consume a protein-restricted diet. Because of limited data in the literature, further research on the optimal intakes and supplementation of micronutrients is required. We performed a systematic review of the relationships between micronutrient levels and clinical outcomes in CKD patients and the effects of different types of RRT on micronutrient levels.

MICRONUTRIENTS IN NON-DIALYSIS CKD PATIENTS

CKD patients can expose to many complications that can be associated with vitamin and trace element disorders. These are increased mortality, increased risk of atherosclerosis, inflammation, oxidative stress, anemia, polyneuropathy, encephalopathy, weakness and frailty, muscle cramps, bone disease, depression, or insomnia. Deficiencies of certain micronutrients are prevalent in non-dialysis CKD patients due to insufficient dietary intake, dietary restrictions, anorexia, proteinuria, unpredictable insufficient absorption, metabolism impairment, and some medications such as diuretics, methotrexate, phosphate binders, ion exchange resins, or immunosuppressants. Table 2 summarizes the plasma or serum levels of vitamins and traces elements in CKD stages 3-5, HD, and PD compared to healthy individuals [6].

Table 2. The changes of levels of vitamins and trace elements in different modalities of renal replacement therapy [6]

Modality	Decreasing levels	Increasing levels	Levels do not change
CKD Stage 3-5	Zn, Se, Mn, B1, vit C	Cu, B6, B9	B3, B12
Hemodialysis	Zn, Se, Mn, B2, B6, B9, vit C	Cu, B1	B1, B2, B3, B9
Peritoneal dialysis	Se, Mn, B2, B6, vit C	B3, B9	Zn, Mn, Cu, B1, B2, B3, B9, B12

Zn: zinc, Se: selenium, Mn: manganese, Cu: copper, B1: thiamine, B2: riboflavin, B3: niacin, B6: pyridoxine, B12: cobalamin, B9: folic acid, vit C: ascorbic acid.

The progression of many complications in CKD patients is associated with oxidative stress. Selenium, iron, and zinc are some of the micronutrients which have roles in the oxidant-antioxidant balance as components of metalloenzymes and participating in reactive oxygen metabolism, free radical scavenging, and hormone activities [7, 8]. The minerals and vitamins exhibit anti-inflammatory and antioxidant activities. An increase in magnesium decreases C-reactive protein. An increase in selenium increases glutathione peroxidase and decreases reactive oxygen species. An increase in vitamin C increases hydroxylase/monooxygenase co-factor, tissue glutathione, and co-antioxidant vitamin E and decreases malondialdehyde, tissue lipid oxidation, and reactive oxygen species. An increase in vitamin E decreases 8-hydroxy-20-deoxyguanosine, lipid peroxidation, nicotinamide adenine dinucleotide phosphate hydrogen activity, and inflammatory mediators. An increase in vitamin D decreases nuclear factor kappa-light-chain-enhancer of activated B cells (NF-kB) signaling pathway, renin-angiotensin-aldosterone system, transforming growth factor-β/Smad and Wnt/β-catenin. An increase in vitamin A decreases the toll-like receptor 4/NF-kB signaling pathway. An increase in vitamin B1 decreases plasma lactate levels [9].

Selenium is an essential micronutrient known as a cofactor in maintaining some antioxidant enzymatic activities such as glutathione peroxidase. Therefore, it has important roles in the protection of cells against destruction by hydrogen peroxide. Selenium and plasma glutathione peroxidases are significantly reduced in CKD patients [7]. An epidemiological study found no significant association between abnormal eGFR with plasma selenium, iron, and zinc levels [10].

Copper has important roles in many biological processes as a component of metalloenzymes. Copper proteins have diverse roles in biological electron transport and oxygen transportation. It can catalyze the formation of reactive oxygen species and cause oxidative damage to cells. Non-protein bound copper ions are toxic by creating reactive oxygen species, which damage biological macromolecules [11]. Serum copper levels are associated with an inflammatory condition. In our study, the most commonly deficient micronutrient was zinc in non-dialysis CKD patients, whereas copper was the least deficient [12]. Shih et al. [13] found no significant difference in copper, iron, and selenium levels at different CKD stages.

Zinc is an essential protein and enzyme component and has an important role in many cellular and subcellular functions, especially anti-inflammatory and antioxidant functions [14, 15]. Many studies on CKD patients showed a zinc plasma deficiency and an abnormal distribution of zinc associated with impairments in immune function, growth, cognitive performance and glucose homeostasis, oxidative stress, and a high burden of atherosclerosis [16-22]. Therefore, imbalances in zinc level play an important role in CKD progression and complications [23]. Several hypotheses may explain zinc deficiency in CKD patients. Low dietary zinc intake, decreased intestinal absorption, increased loss (through the feces or urine), and drug interactions are some of them [24, 25]. Damianaki's study determined lower serum zinc levels and higher 24-hour urinary zinc excretion in non-dialysis CKD patients. They also found positive associations between urinary zinc excretion and CKD status, suggesting that zinc renal handling is altered in CKD [26]. Similarly, Shih et al. [13] found a decrease in CKD patients' zinc concentration from Stages 1 to 4. Zinc has antioxidant roles as a component of superoxide dismutase. Low levels of zinc can cause an increase in the production of hydroxyl radicals. Also, to increase oxidative stress, zinc deficiency increases some interleukins' production and may be associated with atherosclerosis progression in CKD patients [27-29]. Ari et al. [30] demonstrated that the lower the zinc concentration, the higher the carotid artery's intima-media thickness.

Selenium or zinc supplementation in adults with CKD is not routinely recommended as there is little evidence that it improves nutritional, inflammatory, or micronutrient status [31]. Recommended dietary allowance (RDA) is an average daily intake level sufficient to meet the nutrient requirements of nearly all (97%-98%) healthy people. The current RDA for zinc is 8 mg/day for women and 11 mg/day for men in the general population and 55 µg/day for women and men for selenium. It is currently unknown whether a similar intake is recommended at various CKD stages and in the maintenance dialysis population [31].

Manganese is an essential micronutrient that has roles in maintaining homeostasis. The classes of enzymes that require manganese cofactors are broad. The best-known manganese-containing polypeptides are arginase and manganese-containing superoxide dismutase [32]. Nascimento et al. [33] showed that blood manganese levels were significantly inversely associated with kidney function biomarkers. In Shen et al. [34] study, plasma manganese was negatively associated with CKD. In a cross-sectional study, Yang et al. [10] did not found a significant relationship between manganese levels and renal function. Superoxide dismutase activity and lipid peroxidation increase in deficient and excess manganese levels [35]. Therefore, the relationship between manganese and kidney function may negatively and positively correlate and may reflect a U-shaped curve. Potential poisoning risks to humans for high manganese levels were also shown, like zinc [36]. Additional studies are needed to investigate the association between manganese and kidney function.

Vitamin B6 is one of the most deficient micronutrients in non-dialysis CKD patients [12]. However, randomized controlled trials have shown that folate supplementation with or without B-complex vitamins, including vitamins B6 and B12, does not benefit all-cause mortality and/or cardiovascular events in non-dialysis CKD patients [37, 38]. B-complex vitamin supplementation is also not routinely recommended as there is no evidence to suggest a reduction in adverse cardiovascular outcomes. As CKD patients are at risk of vitamin C deficiency, supplements of at least

90 mg/day for men and 75 mg/day for women are recommended to meet adequate intake [31].

The deficiency of essential micronutrients (such as zinc or selenium) and excess of potentially harmful micronutrients are both known to have adverse consequences in the general population [5]. It is plausible that disordered micronutrient status would contribute to morbidity and mortality among CKD patients as well. Unlike dialysis patients, the non-dialysis CKD population remains poorly studied. Therefore, there is a paucity of good-quality evidence to support or oppose routine supplementation on micronutrients in non-dialysis CKD patients.

THE EFFECT OF RENAL REPLACEMENT THERAPY ON SERUM MICRONUTRIENT LEVELS

In the etiology of end-stage renal disease (ESRD), diabetes mellitus (40-45%) is the most common cause, followed by other causes such as hypertension, glomerulonephritis, polycystic kidney disease, and obstructive uropathy. Nowadays, the burden of ESRD is increasing all over the world. Despite optimal treatment, patients with CKD may need RRT (dialysis treatments: home HD, center HD, CAPD or continuous renal replacement therapy [CRRT] and kidney transplantation), especially when the estimated glomerular filtration rate (eGFR) falls below 20 mL/min/1.73 m^2 and/or rapidly worsens to ESRD within 12 months. Waiting for uremic symptoms before starting RRT may increase the patient's risk of malnutrition and mortality. According to USRDS data, HD is the more common dialysis type worldwide. However, the dialysis type preference varies between regions. For example, PD is more common in Hong Kong and the Jalisco region of Mexico, while home HD stands out in New Zealand and Australia [39].

Peritoneal Dialysis

Protein-energy malnutrition is common in PD patients and is related to morbidity and mortality. PD patients are also susceptible to micronutrient (vitamin and mineral) deficiencies due to poor food intake, dialysate and urine losses, intestinal dysfunction, and abnormal metabolism. These deficiencies may exacerbate existing comorbidities (cardiovascular disease, sexual dysfunction, anemia, and sarcopenia) and clinical conditions such as anorexia and decreased taste [40]. Among PD patients, half have been reported to have an inadequate dietary intake of iron, zinc, calcium, vitamin A, B6, C, niacin, and folic acid. Inflammation may both regulate and be regulated by some micronutrients. Patients with inflammation had lower intakes of sodium, calcium, and vitamins A and B2 in this cohort [41]. In dialysis patients' diet, the intake of fruits and vegetables rich in antioxidants (vitamins A, C, and E, trace elements, and other polyphenols) is less. This unbalanced diet also contributes to oxidative stress [42]. Oxidative stress can lead to complications such as peritoneal membrane fibrosis and cardiovascular diseases [43]. Another study revealed that PD patients' intakes of vitamins B1 (15%), B2 (38%), B3 (23%), B6 (39%), folic acid (52%), B12 (32%), E (47%), C (29%), calcium (54%), and zinc (50.5%) with diet and supplements were below the recommended values [44]. Evidence shows that inadequate intake of B-group vitamins, especially vitamins B1 and B3, leads to anorexia [45]. The main causes of vitamin deficiencies in PD patients are losses during peritoneal dialysis and water-soluble vitamins' insufficient intake. PD patients should take 10 mg/day vitamin B6 and at least 1 mg/day folic acid. These amounts in healthy adults are 1.3 to 1.7 mg/day and 400 µg/day, respectively [46]. Vitamin B12 intake in these patients is also low due to insufficient intake of animal protein sources. When a maintenance PD patient using high doses of vitamin C (4 g/day) for several years presented with complaints of decreased vision and arthralgia, proliferative retinopathy, and calcium oxalate crystals due to systemic oxalosis were detected in the ophthalmologic examination. Although vitamin supplements are recommended for these patients due to their deficiencies, uncontrolled high-dose use is also risky [47].

PD patients insufficiently consume meat and dairy products to prevent hyperphosphatemia [45]. The protein restriction probably leads to zinc intake deficiency (5 ± 2 mg/day) in the diet of most PD patients. 54% of PD patients do not take zinc supplements [44]. HD patients are prone to developing zinc deficiency due to removal of zinc by HD, inadequate dietary intake, and reduced gastrointestinal zinc absorption. The prevalence of clinical zinc deficiency (serum level: <60 μg/dL) and subclinical (serum level: 60-80 μg/dL) zinc deficiency was comparable between PD and HD patients (59.6% vs. 70.2% and 40.4% vs. 29.8%, respectively). Multivariate analysis revealed that age, body mass index, and serum albumin level were independent predictors of serum zinc level in both dialysis groups [48]. Zinc deficiency in HD patients may be associated with nonspecific symptoms or conditions such as impaired cognitive function, anorexia, dysgeusia, and erythropoiesis-stimulating agent (ESA) resistant anemia. It has been reported that zinc supplementation can improve appetite, stimulate food intake, increase body mass index, total cholesterol level, and protein catabolism rate in HD patients.

Zinc deficiency may be associated with non-specific symptoms or conditions, which are commonly observed in patients on HD, namely, anorexia, dysgeusia, erythropoiesis-stimulating agent (ESA)-resistant anemia, and impaired cognitive function [49, 50]. It has also been reported that zinc supplementation might improve appetite, stimulate food intake, and increase BMI in patients on HD and that the serum leptin level is decreased in the HD population [51]. In a study evaluating nutritional status by measuring abdominal muscles and fat areas with computed tomography, multiple regression analyzes showed that serum zinc levels were a significant independent predictor of visceral fat areas [52]. Adipose tissues in subcutaneous fat obesity can function normally with the expected release of anti-inflammatory adipokine. In contrast, adipose tissues in visceral fat obesity release increased amounts of pro-inflammatory adipokine, and suppress anti-inflammatory adipokines' secretion, ultimately creating a low-grade inflammation contributes to systemic metabolic and cardiovascular disease [53]. Furthermore, zinc supplementation was shown to increase the total cholesterol level and rate of protein catabolism in patients on HD [54, 55].

Table 3. Comparison of micronutrient levels according to SGA score in CKD patients before RRT

Variables	Normal range	Group 1 (n = 21)	Group 2 (n = 56)	p value
Vitamin B1	33-99 ng/mL	55.4 ± 39.7	67.9 ± 44.6	0.276
Vitamin B6	4.1-43.7 ng/mL	13.6 ± 19.5	34.9 ± 131.5	0.476
Copper	50-155 µg/dL	103.4 ± 20.7	108.4 ± 21.8	0.390
Zinc	70-150 µg/dL	81.6 ± 38.9	76.8 ± 30.9	0.626
Chromium	0.7-28 µg/L	3.7 ± 2.7	6.3 ± 5.6	0.009
Retinol	316-820 µg/L	727.7 ± 481.1	581.3 ± 447	0.234
Selenium	46-143 µg/L	74.4 ± 38.2	108.5 ± 259.9	0.563

The 7-point subjective global assessment (SGA) is a well-validated and reliable tool for malnutrition screening. We divided all patients with CKD in our previous study cohort into two groups according to their SGA scores before RRT: Group 1 (SGA A = well-nourished) and Group 2 (SGA B = moderately malnourished plus SGA C = severely malnourished) [12]. We compared levels of some micronutrients in both groups (unpublished data). Before the start of RRT, chromium levels in Group 1 were significantly higher than Group 2. There was no significant relationship between other micronutrients and SGA scores (Table 3). Serum chromium accumulates in patients receiving dialysis as RRT. The recent study found that the serum chromium level of patients treated with PD was significantly higher than in patients treated with HD (5.0 (3.24-6.15) vs. 1.83 (1.29-2.45) mcg/L). Multivariate analysis also showed a significant relationship between PD modality and serum chromium level (OR: 11.87, 95% CI: 2.85-49.52) [56]. Despite significantly increased liver chromium concentrations in the autopsy of PD patients, the clinical significance of chromium accumulation in chronic dialysis patients remains unclear [57]. This observation has been explained by a few mechanisms seen in PD, unlike HD: continuous contact between dialysate and blood in PD patients, decrease in chromium concentration in peritoneal dialysate after 4 hours (each change adds a significant amount of chromium to the patient's blood), and the transfer of chromium (an easily dialysable form of chromium-lactate) from dialysate to blood, depending on its chemical form [56].

Hemodialysis

Nephrologists routinely recommend salt restriction, but HD patients have poor adherence to a low-salt diet. A recent study evaluated the 3-day dietary intake of 127 chronic hemodialysis patients. The authors found that low daily sodium intake (<1500 mg) was associated with an insufficiently low intake of calories (21.1 ± 6.6 vs. 27.1 ± 10.4 kcal/kg/day, p = 0.0001), protein (0.823 ± 0.275 vs. 1.061 ± 0.419 g/kg/day, p = 0.0003), minerals (phosphorus 749.4 ± 213.4 vs. 910.2 ± 265, p = 0.0004; potassium 1461.1 ± 469 vs. 1779 ± 583 mg, p = 0.001), trace elements (copper 31.5 ± 18.2 vs. 68.6 ± 36.1 mg, p < 0.0001; selenium 42.8 ± 13.4 vs. 68.1 ± 23.7 mg, p < 0.0001; iron 6.56 ± 1.97 vs. 7.81 ± 2.58 mg, p = 0.003; zinc 6.73 ± 2.12 vs. 8.39 ± 2.65 mg, p=0.0003) and vitamin B1 (0.68 ± 0.24 vs. 1.12 ± 0.35 mg, p < 0.0001) compared to those who received ≥1500 mg. Vitamin A (437.7 ± 214.1 vs. 460.5 ± 251.1 µg), B2 (1.02 ± 0.30 vs. 1.13 ± 0.38 mg), B3 (12.3 ± 4.7 vs. 13.8 ± 4.4 mg), C (44.5 ± 30.3 vs. 48.2 ± 47.2 mg) and E (9.1 ± 2.3 vs. 9.5 ± 2.6 mg) intakes were comparable [58]. Therefore, while reducing the daily sodium intake in chronic HD patients, insufficient macro and micronutrient intake may lead to a loss of protein and energy and a deficiency of trace elements and vitamins.

A clinical cross-sectional study examined the association of adequate macronutrient and micronutrient intake with traditional and non-traditional cardiovascular risk factors in 492 HD patients by 3-day diet recording and 24-hour diet recall [59]. The percentage of patients with adequate intake was significantly lower, but cardiovascular disease risk was quite high. The study noted that adequate dietary nutrient intake was associated with an up to 84% lower risk of developing cardiovascular disease, which prevented cardiovascular disease and death. Vitamin B has been shown to improve cardiovascular outcomes in HD patients [60]. However, no effect of vitamin B on cardiovascular disease risk was observed in this study, although a randomized study concluded that folic acid and vitamin B complex significantly reduced homocysteine and hs-CRP levels and increased serum albümin [61].

Cardiovascular diseases are the most common cause of death in uremic patients. Hyperhomocysteinemia is one of the cardiovascular risk factors associated with uremia causing cardiovascular burden. High homocysteine levels promote atherosclerosis through oxidative stress increase, endothelial dysfunction, and thrombosis induction. The epigenetic landscape in patients with CKD and ESRD is associated with the uremic phenotype. Epigenetic alterations linked with DNA methylation (hypo- and hypermethylation) can provide molecular explanations for the complications associated with altered gene expression in CKD. Dietary micronutrients are important for methylome maintenance and epigenetic regulation in the uremic milieu [62]. Methyl donors (folate [vitamin B9], methionine, choline, betaine) and cofactors (vitamin B6 [pyridoxine], vitamin B12 [cobalamin]) regulate homocysteine levels. Vitamin B supplements and dietary intake of methyl donors in uremic patients may provide cardiovascular protection by lowering homocysteine levels [63]. The rate of vitamin B6 deficiency in HD patients is between 24 to 56%. Depending on the dialyzer used, plasma B6 levels decrease by 28 to 48%. The literature consensus is that HD patients are routinely supplemented with 10-50 mg/day B6 (pyridoxine) [64]. Currently, available evidence does not fully support that hyperhomocysteinemia, folic acid, and vitamin B12 changes are reliable cardiovascular disease and cardiovascular mortality risk markers in patients with CKD and ESRD. Folic acid supplementation with or without vitamin B12 is appropriate adjunctive therapy in patients with CKD. Folic acid supplementation in advanced CKD and dialysis patients can also be given after an accurate folate status assessment [65].

Thiamine deficiency has been investigated in non-CKD patients and maintenance dialysis patients, and the results are controversial. Thiamine deficiency is manifested by various clinical signs such as heart failure, peripheral neuropathy, and encephalopathy resulting in Beriberi and Wernicke-Korsakoff syndrome. Since symptomatic beriberi has been reported in CKD patients, thiamine supplementation may be applied [66]. The dialysis population is aging, and sarcopenia and frailty are also more common in these patients, associated with reduced physical function and poor oral intake. The low physical function may be an independent risk factor for thiamine deficiency (blood thiamine concentration <21.3 ng/mL) [67]. Scurvy occurs due to insufficient vitamin C in the body. Since hyperkalemia is a risk factor for dialysis morbidity and mortality, foods containing high potassium and vitamin C are restricted in HD patients. Most of the scurvy symptoms, such as anemia, weakness, and bleeding gums, are usually seen in CKD patients. Vitamin C loss also increases with dialysis. HD patients may need screening for vitamin C deficiency and vitamin C supplements [68]. A large randomized-controlled trial reports that more frequent hemodialysis treatment does not affect circulating plasma vitamin C concentrations and the prevalence of vitamin C deficiency [69]. The results of this study may ignore current concerns regarding the conversion of ascorbate to oxalate and subsequent accumulation in various tissues (including vessel walls leading to accelerated arterial stiffening, if plasma oxalate levels >50 μg/L) resulting from small case series and reports. Supplementation of 60 or 100 mg of vitamin C per day may be a sensible approach in patients who received extended or more frequent dialysis.

An observational study in HD patients measured higher aluminum (14.7 vs. 9.5 μg/L), cobalt (0.57 vs. 0.44 μg/L), nickel (6.2 vs. 1.8 μg/L), strontium (41.1 vs. 32.7 μg/L), molybden (4.5 vs. 1.37 μg/L), cadmium (0.058 vs. 0.025 μg/L) and lead (0.55 vs. 0.30 μg/L) levels, and lower lithium (3.8 vs. 76.2 μg/L), manganese (1.19 vs. 1.66 μg/L), copper (942 vs. 1044 μg/L), zinc (604 vs. 689 μg/L), selenium (71.0 vs. 103.8 μg/L), rubidium (201 vs. 300 μg/L) and barium (0.59 vs. 8.7 μg/L) levels compared to healthy control group [70]. Similarly, a new meta-analysis showed that chronic HD patients had higher cadmium, chromium, copper, lead, and vanadium levels and lower selenium, zinc, and manganese levels compared to healthy controls. Cadmium, chromium, nickel, and vanadium probably accumulate in HD patients. Copper and lead may accumulate in HD patients. Manganese, selenium, and zinc probably are deficient in HD patients [71]. Micronutrients such as manganese are associated with many metalloenzymes and proteins involved in cell metabolism, production of neurotransmitters, and regulatory pathways that control oxidative stress. Subclinical manganism can cause various symptoms, including structural symptoms, behavioral and cognitive dysfunction. Manganese accumulation has been demonstrated in the basal ganglia in HD patients. In a recent study, although blood manganese levels were similar in HD, PD, and kidney transplant patients, only HD patients had symmetrical basal ganglion hyperintensity showing manganese deposition in cranial magnetic resonance images. Moreover, they found a significant relationship between manganese deposition and dialysis duration, suggesting possible contamination with dialysis fluid [72]. Manganese deposition in the brain of maintenance HD patients causes a clinical syndrome with Parkinsonian features. A female patient who had been on irregular dialysis treatment for 1 year due to ESRD secondary to diabetes was investigated for chronic headache and was diagnosed with manganism. Water contamination during dialysis sessions could have caused manganism since she had no history of excessive manganese intake in diet or supplements. The limited treatment options (L-DOPA or chelation with intravenous calcium sodium ethylenediaminetetraacetate) for manganese toxicity failed. Successful kidney transplantation completely resolved manganism clinically and radiologically [73].

A recent study reports the following correlations between observed erythropoiesis stimulating agent (ESA) response variability and different metal concentrations of incident or chronic HD patients. Hemoglobin is negatively associated with cadmium while positively associated with antimony, arsenic, and lead; the amount of ESA dose is negatively correlated with vanadium while positively associated with arsenic. The hemoglobin-dose averaged response of patients to ESAs is most strongly positively associated with arsenic [74].

Selenium is required to synthesize selenoproteins, such as glutathione peroxidase (GPx), in the anti-oxidative defense system. Selenium deficiency has been identified in HD patients due to low food intake, increased urinary-dialytic losses, poor intestinal absorption, abnormal binding of selenium to carrier proteins, and the use of drugs that compromise nutrient absorption in treatment and is associated with increased oxidative stress [75]. Serum selenium levels are significantly lower in HD patients than non-dialysis patients, but the consequences of such deficiency are unknown [76]. A double-blind, randomized placebo-controlled study investigated the efficacy of oral selenium supplementation (200 µg/day for 12 weeks) on 80 chronic HD patients' nutritional status. This study showed that selenium effectively reduced the malnutrition severity in HD patients by alleviating oxidative stress and inflammation. The serum malondialdehyde levels, subjective global assessment, and malnutrition-inflammation scores significantly decreased in the selenium group compared to the placebo group [77]. Reduced selenium levels can contribute to endothelial dysfunction, affect the coronary flow, and promote accelerated atherosclerosis [78]. Besides, serum levels of selenium are inversely associated with the risk of HD patients' mortality [79]. Sleep disturbances are common in HD patients. A recent study found that HD patients with severe sleep disturbances are older, have higher serum parathyroid hormone levels, and lower selenium levels than those who do not experience sleep disorders. However, blood zinc, manganese, copper, and lead levels were comparable in both groups [80]. Animal studies showed that selenium deficiency might predispose to arsenic toxicity [81]. Therefore, HD patients may be at higher than average risk of arsenic toxicity because of low selenium status. However, a recent study did not provide evidence to support cadmium or arsenic toxicity in CKD patients in the analysis of biological samples, including urine, hair, and kidney tissue [82].

In the systematic review and dose-response meta-analysis, subgroup analyzes showed that zinc supplementation increased body weight in HD patients (weighted mean difference = 1.02 kg) [51]. Leptin regulates energy hemostasis and food intake via central nervous system signaling.

Zinc plays a role in modulating leptin, and zinc supplementation may increase leptin synthesis and improve leptin sensitivity. Zinc deficiency is common in HD patients. Some studies have shown that abnormal taste perception and smell acuity, which leads to inadequate food intake and weight loss, improved after zinc supplementation in dialysis patients and energy intake increased. The antioxidant and anti-inflammatory properties of zinc may also benefit [83-85]. Trace element supplements such as oral zinc and selenium increase these elements' blood levels in dialysis patients. Correcting zinc deficiency reduces the risk of infection and all-cause mortality in the general population, but the effect of trace element supplementation on clinical outcomes in HD patients is unclear and limited data are available [71]. Low blood zinc levels were independently associated with persistent intradialytic hypertension in maintenance HD patients. A general linear mixed model showed that each mg/L point decrease in the basal mean blood zinc level increases the peri-dialytic systolic blood pressure change by 4.524 mmHg [86].

Absorption of copper occurs in the stomach and proximal duodenum. The incidence of copper deficiency has increased due to the increase in bariatric surgery. A chronic HD patient who developed severe pancytopenia associated with copper deficiency secondary to malabsorption syndrome after gastric bypass surgery has been reported. Copper deficiency causes pancytopenia, myelopathy and peripheral neuropathy, and osteoporosis [87]. Zinc levels are lower in HD patients compared to normal healthy subjects. An inverse correlation has been shown between zinc and copper levels. Zinc acetate administration may increase the risk of copper deficiency. Zinc deficiency (<60–80 mcg/dL) induces copper deficiency, and copper deficiency causes myeloneuropathy. The upper limit of zinc is 109.7 µg/dL, and the upper safety limit is 78.3 µg/dL to prevent copper deficiency in HD patients. It binds to zinc as an albumin carrier. Serum low zinc concentration is associated with hypoalbuminemia [88].

Continuous Renal Replacement Therapy

Continuous renal replacement therapy (CRRT) is widely used for RRT in the care of critically ill patients with acute and chronic renal dysfunction in intensive care units with hemodynamic instability. CRRT has the advantage of superior volume control compared to intermittent treatment modalities. CRRT can lead to depletion of water-soluble vitamins and trace elements. A recent study revealed that certain vitamins and trace elements are low in critically ill patients requiring CRRT. The deficiency rates were as follows. 16% in thiamine, 67% in pyridoxine, 87% in ascorbic acid, 33% in folic acid, 38% in zinc, and 6% in copper [89]. The general recommendation is to supplement vitamins and trace elements in critically ill patients in intensive care units, taking into account their kidney function [90]. All critically ill patients receive an empirical standard micronutrient supplement. At the beginning of the CRRT, thiamine (100 mg), pyridoxine (200 mg), and folic acid (1 mg) in 50 mL 0.9% sodium chloride infusion are given over 30 minutes. If the patient continues to CRRT, it is repeated every 48 hours [89]. A recent study compared the effects of standard volume hemofiltration (SVHF), high-volume hemofiltration (HVHF), and two new CRRT modes [double filtration hemofiltration (DHF) and dialysate reversible hemodiafiltration (DHDF)] on clearance rates of micronutrients by maintaining a high clearance rate of midsized molecules [91]. The clearance of folic acid, copper, and zinc by HVHF was much greater than that SVHF, DHF, and DHDF. There was no difference among the SVHF, DHF, and DHDF groups. As a result, the loss of water-soluble micronutrients was much greater by HVHF than by SVHF. Two novel CRRT modes (DHF and DHDF) were associated with a substantial reduction in the loss of micronutrients, which was as low as SVHF. Large clinical trials are needed to assess the CRRT systems' effects on the micronutrients and their supplementation requirements in critically ill patients.

Kidney Transplantation

A successful kidney transplant improves the nutrition of patients. Our previous study observed no difference in nutrition habits between male and female kidney recipients by a 3-day dietary regimen. Also, vitamin intakes of both genders were similar (Table 4) [92].

In the Taiwan study, kidney transplant recipients had low intakes of various vitamins such as folic acid or vitamin B6 other than vitamin E due to their higher vitamin E intake in their diet [93]. One study in kidney transplant recipients found the prevalence of vitamin B12 deficiency at 14% due to reduced dietary intake. Vitamin B12 deficiency led to higher adiposity in female recipients. Vitamin B12 deficiency was associated with molecular mechanisms of obesity, insulin resistance, lipogenesis, adipogenesis, and adverse metabolic phenotype. Among recipients with adequate dietary vitamin B12 intake, recipients using MMF had a higher frequency of vitamin B12 deficiency than those using azathioprine. MMF often causes diarrhea, leading to vitamin B12 malabsorption due to villous atrophy in the duodenum and erosive inflammation in the ileum [94]. A meta-analysis evaluated the effects of homocysteine-lowering therapy on cardiovascular mortality in kidney transplant patients with functioning grafts. But meta-analysis was not applied as only a single eligible study (FAVORIT Study 2006) was identified. Although multivitamin supplements effectively reduced homocysteine levels at follow-up, this study found no evidence that the intervention prevented cardiovascular disease and impacted any outcomes, including cardiovascular mortality, all-cause death, myocardial infarction, stroke, the onset of RRT, or all adverse events [95]. Hyperhomocysteinemia and cognitive impairment are common among kidney transplant recipients. The FAVORIT study compared the efficacies of high dose (5 mg folic acid, 1 mg vitamin B12, and 50 mg vitamin B6) and low dose (folate-free 2 µg vitamin B12 and 1.4 mg B6 vitamin) daily multivitamin supplements on homocysteine reduction and cognitive function. Homocysteine levels decreased significantly in both groups, and this effect was more pronounced in the intervention group. Vitamin B supplementation failed to normalize homocysteine in most patients. Ultimately, the high dose of vitamin B supplements provided moderate benefit for some cognitive function in this cohort [96].

Table 4. Comparisons of daily dietary vitamin intakes in in male and female kidney transplant recipients* [92]

	Daily vitamin need in healthy people	Female (n = 33)	Male (n = 29)
Vitamin A	5000 IU	1540.8 ± 96.9	1411.4 ± 96.9
Vitamin E	30 IU	14.3 ± 1.2	14.6 ± 1.01
Vitamin B1	1.5 mg	0.95 ± 0.05	0.95 ± 0.05
Vitamin B2	1.7 mg	1.45 ± 0.07	1.45 ± 0.07
Vitamin B6	2.0 mg	1.36 ± 0.07	1.46 ± 0.07
Folic acid	400 µg	333.5 ± 17.5	322.0 ± 14.2
Vitamin C	60 mg	145.1 ± 7.18	132.2 ± 7.08

* $p > 0.05$.

Due to altered vitamin B6 handling, vitamin B6 deficiency is common after kidney transplantation, and vitamin B6 deficiency status was independently associated with an increased risk of cardiovascular mortality compared with vitamin B6 adequate status [97]. In another study of the same group on this topic, vitamin B6-deficient kidney transplant recipients have worse functional vitamin B6 status than healthy controls and recipients with adequate vitamin B6. A worse functional vitamin B6 status in kidney transplant recipients increases the risk of long-term mortality due to cancer and infectious disease [98]. Vitamin B6 status impacts the body's inflammatory and immune responses. In kidney allograft recipients, an inverse correlation was found between the plasma concentrations of vitamin B6 and the proliferative response of peripheral blood mononuclear cells and cellular immunity markers such as CD28 (+) lymphocyte subsets [99]. A study showed that kidney transplant recipients treated with polyclonal anti-thymocyte globulin (ATG) have lower vitamin B6 stores and active forms. However, this study found no association between vitamins B6, immune response cells, and inflammatory cytokines [100]. Vitamin C depletion also has been associated with an almost two-fold higher risk of kidney transplant recipients' mortality at a median follow-up of 7 years. Vitamin C may have a beneficial effect by reducing inflammation [101].

Serum zinc levels are usually low in patients with CKD, but zinc loss may increase after kidney transplantation, and the deficiency may become clinically apparent. Increased urinary zinc excretion may be responsible for abnormalities in taste acuity and zinc metabolism for continuing up to 1 year after successful kidney transplantation [102]. Even zinc deficiency due to gastrointestinal tuberculosis has developed in one recipient [103]. Kidney allografts with kidney failure accumulate less copper, cadmium, lead, mercury, and vanadium than the kidneys with other pathologically altered kidneys. Immunosuppressive drug usage likely decreases heavy metals concentration in some organs, such as the kidney [104]. Mycophenolate mofetil (MMF), commonly used in kidney transplant recipients, affects trace elements' serum concentrations. High concentrations of arsenic in the blood can lead to bladder, lung, and skin cancers. Unlike the antioxidant trace element selenium, arsenic is involved in oxidative stress leading to renal graft failure. MMF also exhibits antioxidant properties [105]. However, in one study, patients receiving regimens based on MMF, cyclosporine A, and glucocorticoids were reported to have higher arsenic levels than those who received regimens based on MMF, tacrolimus, and glucocorticoids. Male kidney transplant recipients who used MMF also had higher levels of arsenic in red blood cells than those who did not use MMF. Selenium level was significantly higher in younger patients compared to older patients [106]. In another study, the patients treated with MMF had significantly higher plasma zinc and copper concentrations in kidney transplant recipients than the control group [107]. Because it increases zinc excretion, corticosteroid therapy may cause zinc deficiency. MMF may also affect copper balance, but further research and explanation of its mechanism are needed.

If kidney donors use dietary supplements, including micronutrients (vitamins and minerals), the supplements may affect metabolic parameters that mask the future risk of chronic diseases such as diabetes mellitus and hypertension. They do not compromise any aspect of the surgical or perioperative period including anesthesia, analgesia, and bleeding risk. There may be a risk of nephrotoxicity associated with supplementation after the donation [108].

MICRONUTRIENT SUPPLEMENT

DOPPS (Dialysis Results and Application Model Studies) reported that water-soluble vitamin supplements were associated with a significantly lower risk of mortality [109]. A subsequent single-center prospective study also showed that multivitamin supplements significantly reduced the risk of all-cause mortality in ESRD patients compared to those who did not take the supplement [110]. However, no later prospective randomized trial could achieve the same results [6]. In a randomized, double-blind, multicenter study, 650 HD patients were divided into two groups. Patients randomly received orally either 5 mg folic acid, 50 µg cobalamin, and 20 mg vitamin B6 (active treatment) or 0.2 mg folic acid, 4 µg cobalamin, and 1.0 mg vitamin B6 (placebo) after each dialysis session (usually 3 times per week) for an average of 2 years. There was no significant difference between ESRD patients receiving high-dose and low-dose vitamins in total mortality (31% versus 28%) and the frequency of fatal and non-fatal cardiovascular events (25% versus 30%), respectively. Interestingly, homocysteine levels significantly reduced by 35%, and folate levels increased 5-fold in the high-dose group, although there was no significant change in the low-dose group [37]. One center reported that an oral multivitamin supplement containing oral 100 mg thiamine hydrochloride (28-43 mg/day), 20 mg riboflavin (6-9 mg/day), 50 mg pyridoxine hydrochloride (15-21 mg/day), 6 mg folic acid (1.7-2.5 mg/day) and 500 mg ascorbic acid (140-215 mg/day) systematically after dialysis session provided adequate vitamin levels in the long-term in almost all patients using high-flux HD [111]. Oxidative stress is associated with cardiovascular mortality, malnutrition, and inflammation. It begins in the early stages of CKD, gradually increases as GFR decreases, and exacerbates after dialysis. HD patients have increased oxidative stress and inflammation and decreased total antioxidant capacity [112]. Vitamin E-

Coated membranes and many antioxidants, including vitamins B, C, D, and E, and trace elements (zinc, etc.), have been studied to improve dialysis patients' oxidative stress [113].

A group of investigators evaluated the quality of 11 clinical practice guidelines on nutritional management in patients with CKD [114]. The authors reported that the quality of development for most of the guidelines was unsatisfactory. According to the revised Appraisal of Guidelines for Research and Evaluation tool (AGREE II) tool, the best guidelines in terms of quality were the Dietitians Association of Australia (DAA), the European Society for Clinical Nutrition and Metabolism (ESPEN) (enteral nutrition), and the National Kidney foundation-Kidney Disease Outcomes Quality Initiative (KDOQI) [31,115,116]. DAA guidelines recommends that patients with CKD Stage 4 on a restricted protein diet (<0.75 g/kg ideal body weight/day) should receive thiamine (>1 mg/day), riboflavin (1-2 mg/day) and vitamin B6 (1.5-2 mg/day) [115]. ESPEN guidelines on enteral nutrition recommend supplementing water-soluble vitamins (folic acid 1 mg/day, pyridoxine 10-20 mg/day, and vitamin C 30-60 mg/day). Since routine HD does not cause significant trace element losses, supplementation of zinc (15 mg/day) and selenium (50-70 mg/day) may benefit depleted patients. Pyridoxine (10 mg/day) and vitamin C (100 mg/day) supplements may be given to acutely ill CAPD patients [116]. The German parenteral nutrition guidelines for renal failure recommend water-soluble vitamin supplements at approximately twice their normal daily requirement in patients with RRT and malnourished patients without RRT. Vitamin C >250 mg/day is clearly not recommended as it may increase oxalate levels. Vitamin E supplements can be given to patients with chronic kidney failure and accompanying acute diseases and patients with acute renal failure [117]. The EBPG nutrition guidelines recommend thiamine (B1, 1.1-1.2 mg/day), riboflavin (B2, 1.1-1.3 mg/day), pyridoxine (B6, 10 mg/day), ascorbic acid (vitamin C, 75-90 mg/day), folic acid (folate, B9, 1 mg/day), cobalamin (B1, 2.4 mg/day), niacin (B3, nicotinamide, nicotinic acid, vitamin PP, 14-16 mg/day), biotin (B8, 30 mg/day) and pantothenic acid (B5, 5 mg/day) supplements. However, the guideline does not recommend the routine use of vitamin A, vitamin K, zinc, and selenium supplements and advocates daily intake of these micronutrients through the diet (700-900 mg vitamin A, 90-120 mg vitamin K, 8-12 mg for women and 10-15 mg for men zinc, and selenium supplementation). Besides, the guideline recommends the supplements of vitamin E (400-800 IU/day alpha-tocopherol) for in secondary prevention of cardiovascular events and for preventing recurrent muscle cramps, transient vitamin K (10 mg/day) in patients receiving long term antibiotic treatment or those with altered coagulant activity, zinc (50 mg/day for 3-6 months) for a chronic inadequate protein/energy intake and symptoms evoking zinc deficiency, and selenium (3-6 months) in patients with symptoms evoking selenium deficiency [4]. The Renal Nutrition Group (RNG) of the British Dietetic Association committee recommends routine water-soluble vitamin supplements for HD patients. RNG recommends supplementing other micronutrients such as zinc or selenium only if there are consistent symptoms and biochemical evidence of deficiency. It may be possible for modern dialysis treatments to remove more vitamins. However, nutritional evaluation should show that dietary vitamin intake is low or the dialysis dose is unusually high [118].

A recent guideline provided evidence-based recommendations for clinical nutrition in hospitalized patients with acute kidney disease (ACD) and CKD. This guideline recommended monitoring and supplementing water-soluble vitamins and trace elements because of increased requirements during renal failure and critical illness and large effluent losses during RRT. The authors also highlighted more attention to selenium, zinc, copper, vitamin C, folate, and thiamine [119]. The 2020 update to KDOQI Clinical Practice Guidelines for Nutrition in CKD has targeted a large renal population (adults with stages 1-5 CKD who are not receiving dialysis, ESRD including those on dialysis, and individuals with a functional kidney transplant) [120]. Routine folate supplements with or without B-complex are not recommended in adults with CKD stages 3-5D or transplant recipients who had hyperhomocysteinemia due to kidney disease (strong evidence). In adults with CKD stages 1-5D (fair evidence) and transplant recipients (consensus evidence) with CKD, folate, vitamin B12, or B complex supplement is recommended to correct folate or vitamin B12 deficiency/insufficiency based on clinical signs and symptoms. In adults with CKD 1-5D or transplant recipients at risk of vitamin C deficiency, supplementation may be considered (at least 90 mg/day for men and 75 mg/day for women) (consensus evidence). Routine supplement of vitamin A or E is risky in adult patients with CKD 5D on HD or PD due to the potential for vitamin toxicity (consensus evidence). The guideline does not recommend routine supplements of selenium or zinc in adults with CKD 1-5D as there is little evidence that it improves nutritional, inflammatory, or micronutrient status (weak evidence).

Standard multivitamin/multimineral supplements (MVMS) are often prescribed to the healthy population [121]. Nephrologists should be aware that a significant proportion of CKD patients use vitamin and/or trace element supplements without a physician's recommendation. Collaboration with a dietitian experienced in the renal diet will provide a more accurate assessment. The patient should be aware that micronutrient support may benefit in the presence of a deficiency/insufficiency but may also be exposed to the risk of toxicity in unnecessary use. The content of MVMS may not be suitable for use in CKD patients at risk of deficiency. MVMS supplements can be reformulated according to the type of RRT in patients with CKD, taking into account the available evidence.

CONCLUSION

Our previous study compared the effects of 3 different RRT modalities (HD, PD, and kidney transplantation) on different micronutrient levels and nutritional status immediately before and 6 months after RRT in ESRD patients [12]. Our results support the recommendation for supplementing water-soluble vitamins such as vitamins B1 and B6 and minerals such as zinc in patients with ESRD. Kidney transplant appears to be superior to other types of RRT in terms of nutritional parameters. Strict dietary restrictions in HD patients increase the risk of deficiency of micronutrients, especially water-soluble vitamins. Recommendations for vitamin supplementation in HD patients are not concrete, and there is a need for more data on the effect of dialysis and uremic status on water-soluble vitamins. Studies in the field of micronutrients in patients with chronic kidney disease are heterogeneous. They cannot give adequate answers to some issues such as micronutrient intake in the diet, micronutrient support, additional dose, duration of intervention, and outcomes. Therefore, guidelines cannot make accurate and reliable evidence-based recommendations. The 2020 KDOQI guidelines highlight the need for individualization for micronutrient supplements [122]. It is important to investigate micronutrient deficiency along with malnutrition in patients with ESRD. Therefore, there is a need to develop more precise and cheaper methods to measure micronutrient levels. Micronutrient replacement in CKD should be tailored to individual needs.

REFERENCES

[1] Weiner, Daniel E., Hocine Tighiouart, Manish G Amin, Paul C Stark, Bonnie MacLeod, John L Griffith, Deeb N Salem, Andrew S Levey, and Mark J Sarnak. 2004. "Chronic kidney disease as a risk factor for cardiovascular disease and all-cause mortality: A pooled analysis of community-based studies." *Journal of the American Society of Nephrology* May;15(5):1307-15. doi: 10.1097/01.asn.0000123691.46138.e2.

[2] Muscaritoli, Maurizio., Alessio Molfino, Maria Rosa Bollea, and Filippo Rossi Fanelli. 2009. "Malnutrition and wasting in renal disease." *Current Opinion in Clinical Nutrition and Metabolic Care* Jul;12(4):378-83. doi: 10.1097/MCO.0b013e32832c7ae1.

[3] Pazirandeh, Sassan., David L Burns, and Ian J Griffin. Overview of dietary trace elements. *UpToDate 2021*. Last modified Jul 17, 2020. https://www.uptodate.com/contents/overview-of-dietary-trace-elements. Accessed March 15.

[4] Fouque, Denis., Marianne Vennegoor, Piet ter Wee, Christoph Wanner, Ali Basci, Bernard Canaud, Patrick Haage, Klaus Konner, Jeroen Kooman, Alejandro Martin-Malo, Lucianu Pedrini,

Francesco Pizzarelli, James Tattersall, Jan Tordoir, and Raymond Vanholder. 2007. "EBPG guideline on nutrition." *Nephrology Dialysis Transplantation* May;22 Suppl 2:ii45-87. doi: 10.1093/ndt/gfm020.

[5] Shrimpton, Roger., Rainer Gross, Ian Darnton-Hill, and Mark Young. 2005. "Zinc deficiency: what are the most appropriate interventions?" *BMJ* Feb 12;330(7487):347-9. doi: 10.1136/bmj.330.7487.347.

[6] Jankowska, Magdalena., Bolesław Rutkowski, and Alicja Dębska-Ślizień. 2017. "Vitamins and microelement bioavailability in different stages of chronic kidney disease." *Nutrients* Mar 15;9(3):282. doi: 10.3390/nu9030282.

[7] Zachara, Bronislaw A. 2015. "Selenium and selenium-dependent antioxidants in chronic kidney disease." *Advances in Clinical Chemistry* 68:131-51. doi: 10.1016/bs.acc.2014.11.006.

[8] Tapiero, Haim., and Kenneth D Tew. 2003. "Trace elements in human physiology and pathology: zinc and metallothioneins." *Biomedicine and Pharmacotherapy* Nov;57(9):399-411. doi: 10.1016/s0753-3322(03)00081-7.

[9] Rapa, Shara Francesca., Biagio Raffaele Di Iorio, Pietro Campiglia, August Heidland, and Stefania Marzocco. 2019. "Inflammation and oxidative stress in chronic kidney disease-potential therapeutic role of minerals, vitamins and plant-derived metabolites." *International Journal of Molecular Sciences* Dec 30;21(1):263. doi: 10.3390/ijms21010263.

[10] Yang, Fei Yang., Xiping Yi, Jian Guo, Shuaishuai Xu, Yi Xiao, Xiaoyan Huang, Yanying Duan, Dan Luo, Shuiyuan Xiao, Zhijun Huang, Hong Yuan, Meian He, Minxue Shen, and Xiang Chen. 2019. "Association of plasma and urine metals levels with kidney function: a population-based cross-sectional study in China." *Chemosphere* Jul;226:321-328. doi: 10.1016/j.chemosphere.2019.03.171.

[11] Brewer, George J. 2010. "Copper toxicity in the general population." *Clinical Neurophysiology* Apr;121(4):459-60. doi: 10.1016/j.clinph.2009.12.015.

[12] Dizdar, Oguzhan Sıtkı., Abdulmecit Yıldız, Cuma Bulent Gul, Ali Ihsan Gunal, Alparslan Ersoy, and Kursat Gundogan. 2020. "The effect of hemodialysis, peritoneal dialysis and renal transplantation on nutritional status and serum micronutrient levels in patients with end-stage renal disease; Multicenter, 6-month period, longitudinal study." *Journal of Trace Elements in Medicine and Biology* Jul;60:126498. doi: 10.1016/j.jtemb.2020.126498.

[13] Shih Ching-Tang., Ying-Ling Shiu, Chiou-An Chen, Hsin-Yu Lin, Yeou-Lih Huang, and Ching-Chiang Lin. 2012. "Changes in levels of copper, iron, zinc, and selenium in patients at different stages of chronic kidney disease." *Genomic Medicine Biomarkers and Health Sciences* Dec;4(4):128-30. doi: org/10.1016/j.gmbhs.2013.03.001.

[14] Prasad, Ananda S., and Bin Bao. 2019. "Molecular mechanisms of zinc as a pro-antioxidant mediator: clinical therapeutic implications." *Antioxidants (Basel)* Jun 6;8(6):164. doi: 10.3390/antiox8060164.

[15] Plum, Laura M., Lothar Rink, and Hajo Haase. 2010. "The essential toxin: impact of zinc on human health." *International Journal of Environmental Research and Public Health* Apr;7(4):1342-65. doi: 10.3390/ijerph7041342.

[16] Batista, Maria Nazaré., Lílian Cuppari, Lucia de Fátima Campos Pedrosa, Maria das Graças Almeida, José Bruno de Almeida, Anna Cecília Queiroz de Medeiros, and Maria Eugiènia F Canziani. 2006. "Effect of endstage renal disease and diabetes on zinc and copper status." *Biological Trace Element Research* Jul;112(1):1-12. doi: 10.1385/bter:112:1:1.

[17] Dashti-Khavidaki, Simin., Hossein Khalili, Seyedeh-Maryam Vahedi, and Mahboob Lessan-Pezeshki. 2010. "Serum zinc concentrations in patients on maintenance hemodialysis and its relationship with anemia, parathyroid hormone concentrations and

pruritus severity." *Saudi Journal of Kidney Diseases and Transplantation* Jul;21(4):641-5.

[18] Maruyama, Yukio., Akio Nakashima, Akira Fukui, and Takashi Yokoo. 2021. Zinc deficiency: its prevalence and relationship to renal function in Japan." *Clinical and Experimental Nephrology* Mar 17. doi: 10.1007/s10157-021-02046-3.

[19] Mafra, Denise., Lilian Cuppari, and Silvia M F Cozzolino. 2002. "Iron and zinc status of patients with chronic renal failure who are not on dialysis." *Journal of Renal Nutrition* Jan;12(1):38-41. doi: 10.1053/jren.2002.29597.

[20] Fukada, Toshiyuki., Satoru Yamasaki, Keigo Nishida, Masaaki Murakami, and Toshio Hirano. 2011. "Zinc homeostasis and signaling in health and diseases: zinc signaling." *Journal of Biological Inorganic Chemistry* Oct;16(7):1123-34. doi: 10.1007/s00775-011-0797-4.

[21] Kambe, Taihoi, Ayako Hashimoto, and Shigeyuki Fujimoto. 2014. "Current understanding of ZIP and ZnT zinc transporters in human health and diseases." *Cellular and Molecular Life Sciences* Sep;71(17):3281-95. doi: 10.1007/s00018-014-1617-0.

[22] Tamaki, Motoyuki., Yoshio Fujitani, Akemi Hara, Toyoyoshi Uchida, Yoshifumi Tamura, Kageumi Takeno, Minako Kawaguchi, Takahiro Watanabe, Takeshi Ogihara, Ayako Fukunaka, Tomoaki Shimizu, Tomoya Mita, Akio Kanazawa, Mica O Imaizumi, Takaya Abe, Hiroshi Kiyonari, Shintaro Hojyo, Toshiyuki Fukada, Takeshi Kawauchi, Shinya Nagamatsu, Toshio Hirano, Ryuzo Kawamori, and Hirotaka Watada. 2013. "The diabetes-susceptible gene SLC30A8/ZnT8 regulates hepatic insulin clearance." *Journal of Clinical Investigation* Oct;123(10):4513-24. doi: 10.1172/JCI68807.

[23] Choi, Sangyong., Xian Liu, and Zui Pan. 2018. "Zinc deficiency and cellular oxidative stress: prognostic implications in cardiovascular diseases." *Acta Pharmacologica Sinica* Jul;39(7):1120-1132. doi: 10.1038/aps.2018.25.

[24] Pisano, Michele., and Olga Hilas. 2016. "Zinc and taste disturbances in older adults: a review of the literature." *Consultant Pharmacist* May;31(5):267-70. doi: 10.4140/TCP.n.2016.267.

[25] Kambe, Taiho., Tokuji Tsuji, Ayako Hashimoto, and Naoya Itsumura. 2015. "The physiological, biochemical, and molecular roles of zinc transporters in zinc homeostasis and metabolism." *Physiological Reviews* Jul;95(3):749-84. doi: 10.1152/physrev. 00035.2014.

[26] Damianaki, Katerina., Joao Miguel Lourenco, Philippe Braconnier, Jean-Pierre Ghobril, Olivier Devuyst, Michel Burnier, Sebastien Lenglet, Marc Augsburger, Aurelien Thomas, and Menno Pruijm. 2020. "Renal handling of zinc in chronic kidney disease patients and the role of circulating zinc levels in renal function decline." *Nephrology Dialysis Transplantation* Jul 1;35(7):1163-1170. doi: 10.1093/ndt/gfz065.

[27] Shen, Huiyun., Elizabeth Oesterling, Arnold Stromberg, Michal Toborek, Ruth MacDonald, and Bernhard Hennig. 2008. "Zinc deficiency induces vascular pro-inflammatory parameters associated with NF-kappaB and PPAR signaling." *Journal of the American College of Nutrition* Oct;27(5):577-87. doi: 10.1080/07315724. 2008.10719741.

[28] Henning, B., M Toborek, and C J Mcclain. 1996. "Antiatherogenic properties of zinc: implications in endothelial cell metabolism." *Nutrition* Oct;12(10):711-7. doi: 10.1016/s0899-9007(96)00125-6.

[29] Mocchegiani, Eugenio., Laura Costarelli, Robertina Giacconi, Catia Cipriano, Elisa Muti, Silvia Tesei, and Marco Malavolta. 2006. "Nutrient-gene interaction in ageing and successful ageing. A single nutrient (zinc) and some target genes related to inflammatory/ immune response." *Mechanisms of Ageing and Development* Jun;127(6):517-25. doi: 10.1016/j.mad.2006.01.010.

[30] Ari, Elif., Yuksel Kaya, Halit Demir, Ebru Asicioglu, and Sıddık Keskin. 2011. "The correlation of serum trace elements and heavy metals with carotid artery atherosclerosis in maintenance

hemodialysis patients." *Biological Trace Element Research* Dec;144(1-3):351-9. doi: 10.1007/s12011-011-9103-0.

[31] Ikizler, T Alp., Jerrilynn D Burrowes, Laura D Byham-Gray, Katrina L Campbell, Juan-Jesus Carrero, Winnie Chan, Denis Fouque, Allon N Friedman, Sana Ghaddar, D Jordi Goldstein-Fuchs, George A Kaysen, Joel D Kopple, Daniel Teta, Angela Yee-Moon Wang, and Lilian Cuppari. 2020. "KDOQI Clinical Practice Guideline for Nutrition in CKD: 2020 Update." *American Journal of Kidney Diseases* Sep;76(3 Suppl 1):S1-S107. doi: 10.1053/j.ajkd. 2020.05.006.

[32] Culotta, Valeria Cizewski., Mei Yang, and Thomas V O'Halloran. 2006. "Activation of superoxide dismutases: putting the metal to the pedal." *Biochimica et Biophysica Acta* Jul;1763(7):747-58. doi: 10.1016/j.bbamcr.2006.05.003.

[33] Nascimento, Sabrina., Marília Baierle, Gabriela Göethel, Anelise Barth, Natália Brucker, Mariele Charão, Elisa Sauer, Bruna Gauer, Marcelo Dutra Arbo, Louise Altknecht, Márcia Jager, Ana Cristina Garcia Dias, Jerusa Fumagalli de Salles, Tatiana Saint' Pierre, Adriana Gioda, Rafael Moresco, and Solange Cristina Garcia. 2016. "Associations among environmental exposure to manganese, neuropsychological performance, oxidative damage and kidney biomarkers in children." *Environmental Research* May;147:32-43. doi: 10.1016/j.envres.2016.01.035.

[34] Shen, Yu., Zhaoxue Yin, Yuebin Lv, Jiesi Luo, Wenhui Shi, Jianlong Fang, and Xiaoming Shi. 2020 "Plasma element levels and risk of chronic kidney disease in elderly populations ($\geq$ 90 Years old)." *Chemosphere* Sep;254:126809. doi: 10.1016/j.chemosphere. 2020.126809.

[35] Zidenberg-Cherr, S., C L Keen, B Lönnerdal, and L S Hurley. 1983. "Superoxide dismutase activity and lipid peroxidation in the rat: developmental correlations affected by manganese deficiency." *Journal of Nutrition* Dec;113(12):2498-504. doi: 10.1093/jn/113.12. 2498.

[36] Qin, Hai-bo., Jian-ming Zhu, Liang Liang, Ming-shi Wang, and Hui Su. 2013. "The bioavailability of selenium and risk assessment for human selenium poisoning in high-Se areas, China." *Environment International* Feb;52:66-74. doi: 10.1016/j.envint.2012.12.003.

[37] Heinz, Judith., Siegfried Kropf, Ute Domröse, Sabine Westphal, Katrin Borucki, Claus Luley, Klaus H Neumann, and Jutta Dierkes. 2010. "B vitamins and the risk of total mortality and cardiovascular disease in end-stage renal disease: results of a randomized controlled trial." *Circulation* Mar 30;121(12):1432-8. doi: 10.1161/CIRCULATIONAHA.109.904672.

[38] Jamison, Rex L., Pamela Hartigan, James S Kaufman, David S Goldfarb, Stuart R Warren, Peter D Guarino, and J Michael Gaziano, Veterans Affairs Site Investigators. 2007. "Effect of homocysteine lowering on mortality and vascular disease in advanced chronic kidney disease and end-stage renal disease: a randomized controlled trial." *Journal of the American Medical Association* Sep 12;298(10):1163-70. doi: 10.1001/jama.298.10.1163.

[39] Murdeshwar, Himani N., Fatima Anjum. "Hemodialysis." 2020 Dec 4. In: *StatPearls* [Internet]. Treasure Island (FL): StatPearls Publishing; 2021 Jan.

[40] Kiebalo, Thomas., Jacqueline Holotka, Ireneusz Habura, and Krzysztof Pawlaczyk. 2020. "Nutritional status in peritoneal dialysis: Nutritional guidelines, adequacy and the management of malnutrition." *Nutrients* Jun 8;12(6):1715. doi: 10.3390/nu12061715.

[41] Martín-del-Campo, Fabiola., Carolina Batis–Ruvalcaba, Liliana González–Espinoza, Enrique Rojas–Campos, Juan R. Ángel, Norma Ruiz, Juana González, Leonardo Pazarín, and Alfonso M. Cueto–Manzano. 2012. "Dietary micronutrient intake in peritoneal dialysis patients: relationship with nutrition and inflammation status." *Peritoneal Dialysis International* 32:183–191. doi: 10.3747/pdi.2010.00245.

[42] Mekki, Khedidja., Warda Taleb, Nassima Bouzidi, Abbou Kaddous, and Malika Bouchenak. 2010. "Effect of hemodialysis and peritoneal dialysis on redox status in chronic renal failure patients: a comparative study." *Lipids in Health and Disease* Sep 3;9:93. doi: 10.1186/1476-511X-9-93.

[43] H, Noh., Kim JS, Han KH, Lee GT, Song JS, Chung SH, Jeon JS, Ha H, and Lee HB. 2006. "Oxidative stress during peritoneal dialysis: implications in functional and structural changes in the membrane." *Kidney International* Jun;69(11):2022-8. doi: 10.1038/sj.ki.5001506.

[44] As'habi, Atefeh., Iraj Najafi, Hadi Tabibi, and Mehdi Hedayati. 2019. "Dietary intake and its related factors in peritoneal dialysis patients in Tehran, Iran." *Iranian Journal of Kidney Diseases* Jul;13(4):269-276.

[45] ML, Gallagher. 2012. "Intake: the nutrient and their metabolism." In *Krause's Food and the Nutrition Care Process,* 13th ed. edited by Mahan, LK., S Escott-Stump, and JL Raymond, 32-125. Missouri: Elsevier Saunders.

[46] Chazot, C., and JD Kopple. 2013. "Vitamin metabolism and requirements in renal disease and renal failure." In *Nutritional Management of Renal Disease*, 3rd ed. Edited by Kopple, JD., SG Massary, and K Kalantar-Zadeh, 351-76. Boston: Academic Press.

[47] D'Costa, Matthew R., Nelson S. Winkler, Dawn S. Milliner, Suzanne M. Norby, LaTonya J. Hickson, and John C. Lieske. 2019. "Oxalosis associated with high-dose vitamin C ingestion in a peritoneal dialysis patient." *American Journal of Kidney Diseases* Sep;74(3):417-420. doi: 10.1053/j.ajkd.2019.01.022.

[48] Shimizu, Satoshi., Ritsukou Tei, Masahiro Okamura, Nobuteru Takao, Yoshihiro Nakamura, Hidetaka Oguma, Takashi Maruyama, Hiroyuki Takashima, and Masanori Abe. 2020. "Prevalence of zinc deficiency in Japanese patients on peritoneal dialysis: Comparative study in patients on hemodialysis." *Nutrients* Mar 14;12(3):764. doi: 10.3390/nu12030764.

[49] Ortega, RM., A M Requejo, P Andrés, A M López-Sobaler, M E Quintas, M R Redondo, B Navia, and T Rivas. 1997. "Dietary intake and cognitive function in a group of elderly people." *American Journal of Clinical Nutrition* Oct;66(4):803-9. doi: 10.1093/ajcn/66. 4.803.

[50] Kobayashi, Hiroki., Masanori Abe, Kazuyoshi Okada, Ritsukou Tei, Noriaki Maruyama, Fumito Kikuchi, Terumi Higuchi, and Masayoshi Soma. 2015. "Oral zinc supplementation reduces the erythropoietin responsiveness index in patients on hemodialysis." *Nutrients* May 15;7(5):3783-95. doi: 10.3390/nu7053783.

[51] Abdollahi, Shima., Omid Toupchian, Ahmad Jayedi, David Meyre, Vivian Tam, and Sepideh Soltani. 2020. "Zinc supplementation and body weight: A systematic review and dose–response meta-analysis of randomized controlled trials." *Advances in Nutrition* Mar 1;11(2):398-411. doi: 10.1093/advances/nmz084.

[52] Fukasawa, Hirotaka., Hiroki Niwa, Kento Ishibuchi, Mai Kaneko, Takamasa Iwakura, Hideo Yasuda, and Ryuichi Furuya. 2020. "The impact of serum zinc levels on abdominal fat mass in hemodialysis patients." *Nutrients* Feb 28;12(3):656. doi: 10.3390/nu12030656.

[53] Ouchi, Noriyuki., Jennifer L. Parker, Jesse J. Lugus, and Kenneth Walsh. 2011. "Adipokines in inflammation and metabolic disease." *Nature Reviews Immunology* Feb;11(2):85-97. doi: 10.1038/nri2921.

[54] Chevalier, Celia A., George Liepa, Marla D Murphy, Judy Suneson, Anne D Vanbeber, Mary Ann Gorman, and Carolyn Cochran. 2002. "The effects of zinc supplementation on serum zinc and cholesterol concentrations in hemodialysis patients." *Journal of Renal Nutrition* Jul;12(3):183-9. doi: 10.1053/jren.2002.33515.

[55] Jern, Nancy A., Anne D. VanBeber, Mary Anne Gorman, Cynthia G. Weber, George U. Liepa, and Carolyn C. Cochran. 2000. "The effects of zinc supplementation on serum zinc concentration and protein catabolic rate in hemodialysis patients." *Journal of Renal Nutrition* Jul;10(3):148-53. doi: 10.1053/jren.2000.7413.

[56] Feldman, Leonid., Ilia Beberashvili, Ramzia Abu Hamad, Iris Yakov-Hai, Elena Abramov, Walter Wasser, Oleg Gorelik, Roza

Rozenberg, and Shai Efrati. 2019. "Serum chromium levels are higher in peritoneal dialysis than in hemodialysis patients." *Peritoneal Dialysis International* Jul-Aug;39(4):330-334. doi: 10.3747/pdi.2018.00183.

[57] Borguet, F., B Wallaeys, R Cornelis, and N Lameire. 1996. "Transperitoneal absorption and kinetics of chromium in the continuous ambulatory peritoneal dialysis patient--an experimental and mathematical analysis." *Nephron* 72(2):163-70. doi: 10.1159/000188836.

[58] Bossola, Maurizio., Enrico Di Stasio, Antonella Viola, Stefano Cenerelli, Alessandra Leo, Stefano Santarelli, and Tania Monteburini. 2020. "Dietary daily sodium intake lower than 1500 mg is associated with inadequately low intake of calorie, protein, iron, zinc and vitamin B1 in patients on chronic hemodialysis." *Nutrients* Jan 19;12(1):260. doi: 10.3390/nu12010260.

[59] Duong, Tuyen Van., Te-Chih Wong, Chien-Tien Su, Hsi-Hsien Chen, Tzen-Wen Chen, Tso-Hsiao Chen, Yung-Ho Hsu, Sheng-Jeng Peng, Ko-Lin Kuo, Hsiang-Chung Liu, En-Tsu Lin, and Shwu-Huey Yang. 2018. "Associations of dietary macronutrients and micronutrients with the traditional and nontraditional risk factors for cardiovascular disease among hemodialysis patients: A clinical cross-sectional study." *Medicine (Baltimore)* Jun;97(26):e11306. doi: 10.1097/MD.0000000000011306.

[60] Qin, Xianhui., Yong Huo, Craig B Langman, Fanfan Hou, Yundai Chen, Debora Matossian, Xiping Xu, and Xiaobin Wang. 2011. "Folic acid therapy and cardiovascular disease in ESRD or advanced chronic kidney disease: a meta-analysis." *Clinical Journal of the American Society of Nephrology* Mar;6(3):482-8. doi: 10.2215/CJN.05310610.

[61] Chang, Tzu-Yuan., Kang-Ju Chou, Chin-Feng Tseng, Hsiao-Min Chung, Hua-Chang Fang, Yao-Min Hung, Ming-Jei Wu, Huey-Ming Tzeng, Chang-Chung Lind, and Kuo-Cheng Lu. 2007. "Effects of folic acid and vitamin B complex on serum C-reactive protein and albumin levels in stable hemodialysis patients." *Current*

Medical Research and Opinion Aug;23(8):1879-86. doi: 10.1185/ 030079907X218077.

[62] Mafra, Denise., Marta Esgalhado, Natalia A Borges, Ludmila F M F Cardozo, Milena B Stockler-Pinto, Hannah Craven, Sarah J Buchanan, Bengt Lindholm, Peter Stenvinkel, and Paul G Shiels. 2019. "Methyl donor nutrients in chronic kidney disease: Impact on the epigenetic landscape." *Journal of Nutrition* Mar 1;149(3):372-380. doi: 10.1093/jn/nxy289.

[63] Cappuccilli, Maria., Camilla Bergamini, Floriana A Giacomelli, Giuseppe Cianciolo, Gabriele Donati, Diletta Conte, Teresa Natali, Gaetano La Manna, and Irene Capelli. 2020. "Vitamin B supplementation and nutritional intake of methyl donors in patients with chronic kidney disease: A critical review of the impact on epigenetic machinery." *Nutrients* Apr 27;12(5):1234. doi: 10.3390/nu12051234.

[64] Corken, Melissa., and Judi Porter. 2011. "Is vitamin B(6) deficiency an under-recognized risk in patients receiving haemodialysis? A systematic review: 2000-2010." *Nephrology (Carlton)* Sep;16(7): 619-25. doi: 10.1111/j.1440-1797.2011.01479.x.

[65] Capelli, Irene., Giuseppe Cianciolo, Lorenzo Gasperoni, Fulvia Zappulo, Francesco Tondolo, Maria Cappuccilli, and Gaetano La Manna. 2019. "Folic acid and vitamin B12 administration in CKD, why not?" *Nutrients* Feb 13;11(2):383. doi: 10.3390/nu11020383.

[66] Kittanamongkolchai, Wonngarm., Napat Leeaphorn, Narat Srivali, and Wisit Cheungpasitporn. 2013. "Beriberi in a dialysis patient: do we need more thiamine?" *American Journal of Emergency Medicine* Apr;31(4):753. doi: 10.1016/j.ajem.2013.01.013.

[67] Saka, Yosuke., Tomohiko Naruse, Akihisa Kato, Naoto Tawada, Yuhei Noda, Tetsushi Mimura, and Yuzo Watanabe. 2018. "Thiamine status in end-stage chronic kidney disease patients: a single-center study." *International Urology and Nephrology* Oct;50(10):1913-1918. doi: 10.1007/s11255-018-1974-y.

[68] Panchal, Sarju., Christine Schneider, and Kunal Malhotra. 2018. "Scurvy in a hemodialysis patient. Rare or ignored?" *Hemodialysis*

International Oct;22(S2):S83-S87. doi: 10.1111/hdi.12705. Epub 2018 Nov 8.

[69] Raimann, Jochen G., Samer R Abbas, Li Liu, Brett Larive, Gerald Beck, Peter Kotanko, Nathan W Levin, and Garry Handelman; FHN Trial. 2019. "The effect of increased frequency of hemodialysis on vitamin C concentrations: an ancillary study of the randomized Frequent Hemodialysis Network (FHN) daily trial." *BMC Nephrology* May 17;20(1):179. doi: 10.1186/s12882-019-1311-4.

[70] Almeida, Agostinho., Katarzyna Gajewska, Mary Duro, Félix Costa, and Edgar Pinto. 2020. "Trace element imbalances in patients undergoing chronic hemodialysis therapy - Report of an observational study in a cohort of Portuguese patients." *Journal of Trace Elements in Medicine and Biology* Dec;62:126580. doi: 10.1016/j.jtemb.2020.126580.

[71] Tonelli, Marcello., Natasha Wiebe, Brenda Hemmelgarn, Scott Klarenbach, Catherine Field, Braden Manns, Ravi Thadhani, and John Gill; Alberta Kidney Disease Network. 2009. "Trace elements in hemodialysis patients: a systematic review and meta-analysis." *BMC Medicine* May 19;7:25. doi: 10.1186/1741-7015-7-25.

[72] Akcan, Esra., Sultan Özkurt, Garip Sahin, Ahmet Ugur Yalcin, and Baki Adapinar. 2018. "The relation between brain MRI findings and blood manganese levels in renal transplantation, hemodialysis, and peritoneal dialysis patients." *International Urology and Nephrology* Jan;50(1):173-177. doi: 10.1007/s11255-017-1731-7.

[73] George, Pratish., Suceena Alexander, Santosh Varughese, Binita Riya Chacko, Vinoi George David, Venumadhav Gowrugari, Veerasamy Tamilarasi, and Chakko Korula Jacob. 2013. "First case on successful management of manganism with renal transplantation." *Transplantation* May 15;95(9):e58-9. doi: 10.1097/TP.0b013e31828b43e3.

[74] Brier, Michael E., Jessica R Gooding, James M Harrington, Jason P Burgess, Susan L McRitchie, Xiaolan Zhang, Brad H Rovin, Jon B Klein, Jonathan Himmelfarb, Susan J Sumner, and Michael L Merchant. 2020. "Serum trace metal association with response to

erythropoiesis stimulating agents in incident and prevalent hemodialysis patients." *Scientific Reports* Nov 19;10(1):20202. doi: 10.1038/s41598-020-77311-8.

[75] Barroso, Christielle Félix., Liliane Viana Pires, Larissa Bezerra Santos, Gilberto Simeone Henriques, Priscila Pereira Pessoa, Gueyhsa Nobre de Araújo, Camilla Oliveira Duarte de Araújo, Cláudia Maria Costa Oliveira, and Carla Soraya Costa Maia. 2020. "Selenium nutritional status and glutathione peroxidase activity and its relationship with hemodialysis time in individuals living in a Brazilian region with selenium-rich soil." *Biological Trace Element Research* Sep 22. doi: 10.1007/s12011-020-02388-1.

[76] Thompson, Stephanie., and Marcello Tonelli. 2013. "Selenium for malnutrition in hemodialysis patients: have we considered all of the elements?" *Nephrology Dialysis Transplantation* Mar;28(3):498-500. doi: 10.1093/ndt/gfs286.

[77] Salehi, Moosa., Zahra Sohrabi, Maryam Ekramzadeh, Mohammad Kazem Fallahzadeh, Maryam Ayatollahi, Bita Geramizadeh, Jafar Hassanzadeh, and Mohammad Mahdi Sagheb. 2013. "Selenium supplementation improves the nutritional status of hemodialysis patients: a randomized, double-blind, placebo-controlled trial." *Nephrology Dialysis Transplantation* Mar;28(3):716-23. doi: 10.1093/ndt/gfs170.

[78] Bacci, Marcelo R., Lívia S S Cabral, Glaucia L da Veiga, Beatriz da C A Alves, Neif Murad, and Fernando L A Fonseca. 2020. "The impact of inflammatory profile on selenium levels in hemodialysis patients." *Antiinflamm Antiallergy Agents Med Chem* 19(1):42-49. doi: 10.2174/1871523018666190121165902.

[79] Fujishima, Yosuke., Masaki Ohsawa, Kazuyoshi Itai, Karen Kato, Kozo Tanno, Tanvir Chowdhury Turin, Toshiyuki Onoda, Shigeatsu Endo, Akira Okayama, and Tomoaki Fujioka. 2011. "Serum selenium levels are inversely associated with death risk among hemodialysis patients." *Nephrology Dialysis Transplantation* Oct;26(10):3331-8. doi: 10.1093/ndt/gfq859.

[80] Xu, Shilin., De'e Zou, Ruiying Tang, Shuting Li, Wenxuan Chen, Luona Wen, Yun Liu, Yan Liu, and Xiaoshi Zhong. 2021. "Levels of trace blood elements associated with severe sleep disturbance in maintenance hemodialysis patients." *Sleep and Breathing* Mar 5. doi: 10.1007/s11325-021-02336-w.

[81] Biswas, S., G Talukder, and A Sharma. 1999. "Prevention of cytotoxic effects of arsenic by short-term dietary supplementation with selenium in mice in vivo." *Mutation Research* Apr 26;441(1):155-60. doi: 10.1016/s1383-5718(99)00028-5.

[82] Nanayakkara, Shanika., S T M L D Senevirathna, Kouji H Harada, Rohana Chandrajith, Toshiaki Hitomi, Tilak Abeysekera, Eri Muso, Takao Watanabe, and Akio Koizumi. 2019. "Systematic evaluation of exposure to trace elements and minerals in patients with chronic kidney disease of uncertain etiology (CKDu) in Sri Lanka." *Journal of Trace Elements in Medicine and Biology* Jul;54:206-213. doi: 10.1016/j.jtemb.2019.04.019.

[83] Manley, Karen J., Rivkeh Y. Haryono, and Russel S.J. Keast. 2012. "Taste changes and saliva composition in chronic kidney disease." *Renal Society of Australasia Journal* 8(2):56-60.

[84] Mahajan, SK., A S Prasad, J Lambujon, A A Abbasi, W A Briggs, and F D McDonald. 1980. "Improvement of uremic hypogeusia by zinc: a double-blind study." *American Journal of Clinical Nutrition* Jul;33(7):1517-21. doi: 10.1093/ajcn/33.7.1517.

[85] Sprenger, KB., D Bundschu, K Lewis, B Spohn, J Schmitz, and H E Franz. 1983. "Improvement of uremic neuropathy and hypogeusia by dialysate zinc supplementation: a double-blind study." *Kidney International Suppl* Dec;16:S315-8.

[86] Liu, Yun., Yuanyuan Zheng, Liangtao Wang, Xiaoshi Zhong, Danping Qin, Wenxuan Chen, Rongshao Tan, and Yan Liu. 2020. "Lower levels of blood zinc associated with intradialytic hypertension in maintenance hemodialysis patients." *Biological Trace Element Research* Sep 15. doi: 10.1007/s12011-020-02385-4.

[87] Melero, Eva López., Gloria Ruíz-Roso, Ignacio Botella, Sofía Ortego Pérez, María Delgado, and Milagros Fernández Lucas. 2019.

"Pancytopenia due to copper deficiency in a hemodialysis patient." *Nefrologia* Jul-Aug;39(4):451-452. doi: 10.1016/j.nefro.2018.10.011.

[88] Nishime, Keizo., Morihiro Kondo, Kazuhiro Saito, Hisashi Miyawaki, and Takahiko Nakagawa. 2020. "Zinc burden evokes copper deficiency in the hypoalbuminemic hemodialysis patients." *Nutrients* Feb 23;12(2):577. doi: 10.3390/nu12020577.

[89] Kamel, Amir Y., Nisha J. Dave, Vivian M. Zhao, Daniel P. Griffith, Michael J. Connor Jr, and Thomas R. Ziegler. 2018. "Micronutrient alterations during continuous renal replacement therapy in critically ill adults: A retrospective study." *Nutrition in Clinical Practice* Jun;33(3):439-446. doi: 10.1177/0884533617716618.

[90] McClave, Stephen A., Beth E Taylor, Robert G Martindale, Malissa M Warren, Debbie R Johnson, Carol Braunschweig, Mary S McCarthy, Evangelia Davanos, Todd W Rice, Gail A Cresci, Jane M Gervasio, Gordon S Sacks, Pamela R Roberts, and Charlene Compher, Society of Critical Care Medicine; American Society for Parenteral and Enteral Nutrition. 2016. "Guidelines for the Provision and Assessment of Nutrition Support Therapy in the Adult Critically Ill Patient: Society of Critical Care Medicine (SCCM) and American Society for Parenteral and Enteral Nutrition (A.S.P.E.N.)." *Journal of Parenteral and Enteral Nutrition* Feb;40(2):159-211. doi: 10.1177/0148607115621863.

[91] Wu, Buyun., Daxi Ji, Bin Xu, Rong Fan, and Dehua Gong. 2019. "New modes of continuous renal replacement therapy using a refiltering technique to reduce micronutrient loss." *Hemodialysis International* Apr;23(2):181-188. doi: 10.1111/hdi.12709.

[92] Ersoy, Alparslan., Nizameddin Koca, Tuba Gullu Koca, and Canan Ersoy. 2017. "Anthropometric measurements, nutrition and exercise habits in kidney transplant recipients." *Paper presented at 19th European Congress of Endocrinology*, Lisbon, Portugal, May 20-23. Published at 2017-05-03, Endocrine Abstracts 49 EP759. doi:10.1530/endoabs.49.EP759.

[93] Lin, I-Hsin., Te-Chih Wong, Shih-Wei Nien, Yu-Ting Chou, Yang-Jen Chiang, Hsu-Han Wang, and Shwu-Huey Yang. 2019. "Dietary compliance among renal transplant recipients: A single-center study in Taiwan." *Transplantation Proceedings* Jun;51(5):1325-1330. doi: 10.1016/j.transproceed.2019.02.026.

[94] Pontes, Karine Scanci da Silva., Márcia Regina Simas Torres Klein, Mariana Silva da Costa, Kelli Trindade de Carvalho Rosina, Ana Paula Medeiros Menna Barreto, Maria Inês Barreto Silva, and Suzimar da Silveira Rioja. 2019. "Vitamin B(12) status in kidney transplant recipients: association with dietary intake, body adiposity and immunosuppression." *British Journal of Nutrition* Aug 28;122(4):450-458. doi: 10.1017/S0007114519001417.

[95] Kang, Amy., Sagar U Nigwekar, Vlado Perkovic, Satyarth Kulshrestha, Sophia Zoungas, Sankar D Navaneethan, Alan Cass, Martin P Gallagher, Toshiharu Ninomiya, Giovanni F M Strippoli, and Meg J Jardine. 2015. "Interventions for lowering plasma homocysteine levels in kidney transplant recipients." *Cochrane Database of Systematic Reviews* May 4;(5):CD007910. doi: 10.1002/14651858.CD007910.pub2.

[96] Scott, T. M., G Rogers, D E Weiner, K Livingston, J Selhub, P F Jacques, I H Rosenberg, and A M Troen. 2017. "B-vitamin therapy for kidney transplant recipients lowers homocysteine and improves selective cognitive outcomes in the randomized FAVORIT ancillary cognitive trial." *Journal of Prevention of Alzheimer's Disease* 4(3):174-182. doi: 10.14283/jpad.2017.15.

[97] Minović, Isidor., Ineke J Riphagen, Else van den Berg, Jenny E Kootstra-Ros, Martijn van Faassen, Antonio W Gomes Neto, Johanna M Geleijnse, Reinold Ob Gans, Manfred Eggersdorfer, Gerjan J Navis, Ido P Kema, and Stephan Jl Bakker. 2017. "Vitamin B-6 deficiency is common and associated with poor long-term outcome in renal transplant recipients." *American Journal of Clinical Nutrition* Jun;105(6):1344-1350. doi: 10.3945/ajcn.116.151431.

[98] Minović, Isidor., Anna van der Veen, Martijn van Faassen, Ineke J Riphagen, Else van den Berg, Claude van der Ley, António W Gomes-Neto, Johanna M Geleijnse, Manfred Eggersdorfer, Gerjan J Navis, Ido P Kema, and Stephan Jl Bakker. 2017. "Functional vitamin B-6 status and long-term mortality in renal transplant recipients." *American Journal of Clinical Nutrition* Dec;106(6): 1366-1374. doi: 10.3945/ajcn.117.164012.

[99] Jankowska, Magdalena., Marcin Marszałł, Alicja Dębska-Ślizień, Juan J Carrero, Bengt Lindholm, Wojciech Czarnowski, Bolesław Rutkowski, and Piotr Trzonkowski. 2013. "Vitamin B6 and the immunity in kidney transplant recipients." *Journal of Renal Nutrition* Jan;23(1):57-64. doi: 10.1053/j.jrn.2012.01.023.

[100] Jankowska, M., P Trzonkowski, A Dębska-Ślizień, M Marszałł, and B Rutkowski. 2014. "Vitamin B6 status, immune response and inflammation markers in kidney transplant recipients treated with polyclonal anti-thymocyte globulin." *Transplantation Proceedings* Oct;46(8):2631-5. doi: 10.1016/j.transproceed.2014.08.009.

[101] Sotomayor, Camilo G., Michele F Eisenga, Antonio W Gomes Neto, Akin Ozyilmaz, Rijk O B Gans, Wilhelmina H A de Jong, Dorien M Zelle, Stefan P Berger, Carlo A J M Gaillard, Gerjan J Navis, and Stephan J L Bakker. 2017. "Vitamin C depletion and all-cause mortality in renal transplant recipients." *Nutrients* Jun 2;9(6):568. doi: 10.3390/nu9060568.

[102] Mahajan, S. K., J Abraham, S D Migdal, D K Abu-Hamdan, and F D McDonald. 1984. "Effect of renal transplantation on zinc metabolism and taste acuity in uremia. A prospective study." *Transplantation* Dec;38(6):599-602. doi: 10.1097/00007890-19841 2000-00010.

[103] Ghuge, Priyanka., Rusina Karia, and Ram H Malkani. 2018. "Acquired zinc deficiency in a renal transplant recipient with gastrointestinal tuberculosis responding promptly to oral correction." *Saudi Journal of Kidney Diseases and Transplantation* Sep-Oct;29(5):1199-1202. doi: 10.4103/1319-2442.243962.

[104] Wilk, Aleksandra., Barbara Wiszniewska, Anna Rzuchowska, Maciej Romanowski, Jacek Różański, Marcin Słojewski, Kazimierz Ciechanowski, and Elżbieta Kalisińska. 2019. "Comparison of copper concentration between rejected renal grafts and cancerous kidneys." *Biological Trace Element Research* Oct;191(2):300-305. doi: 10.1007/s12011-018-1621-6.

[105] Ferjani, Hanen., Amira El Arem, Aicha Bouraoui, Abedellatif Achour, Salwa Abid, Hassen Bacha, and Imen Boussema-Ayed. 2016. "Protective effect of mycophenolate mofetil against nephrotoxicity and hepatotoxicity induced by tacrolimus in Wistar rats." *Journal of Physiology and Biochemistry* Jun;72(2):133-44. doi: 10.1007/s13105-015-0451-7.

[106] Wilk, Aleksandra., and Barbara Wiszniewska. 2020. "Arsenic and selenium profile in erythrocytes of renal transplant recipients." *Biological Trace Element Research* Oct;197(2):421-430. doi: 10.1007/s12011-019-02021-w.

[107] Kamińska, Jolanta., Joanna Sobiak, Joanna Maria Suliburska, Grażyna Duda, Maciej Głyda, Zbigniew Krejpcio, and Maria Chrzanowska. 2012. "Effect of mycophenolate mofetil on plasma bioelements in renal transplant recipients." *Biological Trace Element Research* Feb;145(2):136-43. doi: 10.1007/s12011-011-9178-7.

[108] Leonberg-Yoo, Amanda K., David Johnson, Nicole Persun, Jehan Bahrainwala, Peter P Reese, Ali Naji, and Jennifer Trofe-Clark. 2020. "Use of dietary supplements in living kidney donors: A Critical Review." *American Journal of Kidney Diseases* Dec;76(6):851-860. doi: 10.1053/j.ajkd.2020.03.030.

[109] Fissell, Rachel B., Jennifer L Bragg-Gresham, Brenda W Gillespie, David A Goodkin, Juergen Bommer, Akira Saito, Takashi Akiba, Friedrich K Port, and Eric W Young. 2004. "International variation in vitamin prescription and association with mortality in the Dialysis Outcomes and Practice Patterns Study (DOPPS)." *American Journal of Kidney Diseases* Aug;44(2):293-9. doi: 10.1053/j.ajkd.2004.04.047.

[110] Domröse, U., J Heinz, S Westphal, C Luley, K.H. Neumann, and J Dierkes. 2007. "Vitamins are associated with survival in patients with end-stage renal disease: a 4-year prospective study." *Clinical Nephrology* Apr;67(4):221-9. doi: 10.5414/cnp67221.

[111] Descombes, E., O Boulat, F Perriard, and G Fellay. 2000. "Water-soluble vitamin levels in patients undergoing high-flux hemodialysis and receiving long-term oral postdialysis vitamin supplementation." *Artificial Organs* Oct;24(10):773-8. doi: 10.1046/j.1525-1594.2000.06553.x.

[112] Senol, Emel., Alpaslan Ersoy, Selda Erdinc, Emre Sarandol, and Mustafa Yurtkuran. 2008. "Oxidative stress and ferritin levels in haemodialysis patients." *Nephrology Dialysis Transplantation* Feb;23(2):665-72. doi: 10.1093/ndt/gfm588.

[113] Liakopoulos, Vassilios., Stefanos Roumeliotis, Andreas Bozikas, Theodoros Eleftheriadis, and Evangelia Dounousi. 2019. "Antioxidant supplementation in renal replacement therapy patients: Is there evidence?" *Oxidative Medicine and Cellular Longevity* Jan 15;2019:9109473. doi: 10.1155/2019/9109473.

[114] Bakaloudi, Dimitra Rafailia., Lydia Chrysoula, Kalliopi Anna Poulia, Evangelia Dounousi, Vassilios Liakopoulos, and Michail Chourdakis. 2021. "AGREEing on nutritional management of patients with CKD-A quality appraisal of the available guidelines." *Nutrients* Feb 15;13(2):624. doi: 10.3390/nu13020624.

[115] Ash, Susan., Katrina Campbell, Helen MacLaughlin, Ellen McCoy, Maria Chan, Kathryn Anderson, Karen Corke, Ruth Dumont, Lyn Lloyd, Anthony Meade, Robyn Montgomery-Johnson, Tracey Tasker, Paulett Thrift, and Bernadeen Trotter. 2006. "Evidence based practice guidelines for the nutritional management of chronic kidney disease." *Nutrition and Dietetics* Sep;63(2):S33-S45. doi: org/10.1111/j.1747-0080.2006.00100.x.

[116] Cano, N., E Fiaccadori, P Tesinsky, G Toigo, and W Druml; DGEM (German Society for Nutritional Medicine), Kuhlmann M, Mann H, Hörl WH; ESPEN (European Society for Parenteral and Enteral Nutrition). 2006. "ESPEN guidelines on enteral nutrition: Adult

renal failure." *Clinical Nutrition* Apr;25(2):295-310. doi: 10.1016/j.clnu.2006.01.023.

[117] Druml, W., and H.P. Kierdorf; Working group for developing the guidelines for parenteral nutrition of The German Association for Nutritional Medicine. 2009. "Parenteral nutrition in patients with renal failure - Guidelines on Parenteral Nutrition, Chapter 17." *German Medical Science* Nov 18;7:Doc11. doi: 10.3205/000070.

[118] Wright, Mark., Elizabeth Southcott, Helen MacLaughlin, Stuart Wineberg. 2019. "Clinical practice guideline on undernutrition in chronic kidney disease." *BMC Nephrology* Oct 16;20(1):370. doi: 10.1186/s12882-019-1530-8.

[119] Fiaccadori, Enrico., Alice Sabatino, Rocco Barazzoni, Juan Jesus Carrero, Adamasco Cupisti, Elisabeth De Waele, Joop Jonckheer, Pierre Singer, and Cristina Cuerda. 2021. "ESPEN guideline on clinical nutrition in hospitalized patients with acute or chronic kidney disease." *Clinical Nutrition* Feb 9;S0261-5614(21)00052-2. doi: 10.1016/j.clnu.2021.01.028.

[120] Handu, Deepa., Mary Rozga, and Alison Steiber. 2020. "Executive Summary of the 2020 Academy of Nutrition and Dietetics and National Kidney Foundation Clinical Practice Guideline for Nutrition in CKD." *Journal of the Academy of Nutrition and Dietetics* Nov 3:S2212-2672(20)31234-X. doi: 10.1016/j.jand.2020.08.092.

[121] Blumberg, Jeffrey B., Hellas Cena, Susan I Barr, Hans Konrad Biesalski, Ricardo Uauy Dagach, Brendan Delaney, Balz Frei, Manuel Ignacio Moreno González, Nahla Hwalla, Ronette Lategan-Potgieter, Helene McNulty, Jolieke C van der Pols, Pattanee Winichagoon, and Duo Li. 2018. "The use of multivitamin/multimineral supplements: A modified Delphi consensus panel report." *Clinical Therapeutics* Apr;40(4):640-657. doi: 10.1016/j.clinthera.2018.02.014.

[122] Ikizler, Talat Alp., and Lilian Cuppari. 2021. "The 2020 Updated KDOQI Clinical Practice Guidelines for Nutrition in Chronic

Kidney Disease." *Blood Purification* Mar 2:1-5. doi: 10.1159/ 000513698.

BIOGRAPHICAL SKETCH

Oguzhan Sıtkı Dizdar

Affiliation: Kayseri City Training and Research Hospital, Department of Internal Medicine and Clinical Nutrition.

Education:

- September 2018 Certificate of Teach the Teachers for Life Long Learning (T-LLL) programme of ESPEN.
- September 2016- At present Molecular Genetic PhD student in Department of Medical Biology, Faculty of Medicine, Erciyes University, Kayseri, Turkey.
- 2015 Department of Clinical Nutrition, Erciyes University Medical Faculty.
- November 2006- March 2012 Residency in Internal Medicine, Uludag University, School of Medicine, Bursa, Turkey.
- September 1998- August 2005 Doctor of Medicine (M.D.) training, Erciyes University, School of Medicine, Kayseri, Turkey.

Research and Professional Experience: Internal medicine, diabetes mellitus, chronic kidney disease, clinical nutrition, micronutrients.

Professional Appointments:

- October 2017 – At present Associate Professor of Medicine, Department of Internal Medicine, Kayseri Training and Research Hospital, Kayseri, Turkey

- September 2016 – At present PhD student, Department of Medical Biology, Erciyes University Medical Faculty, Kayseri, Turkey
- October 2015- October 2017 Senior research fellow, Department of Internal Medicine, Kayseri Training and Research Hospital, Kayseri, Turkey
- September 2015- At present Clinical Nutrition Unit Chief, Kayseri Training and Research Hospital, Kayseri, Turkey
- July 2014- October 2015 Internal Medicine Specialist, Department of Internal Medicine, Kayseri Training and Research Hospital, Kayseri, Turkey
- May 2012- July 2014 Internal Medicine Specialist, Afsin State Hospital Ministry of Health, Kahramanmaras, Turkey

Honors: The Scientific and Technological Research Council of Turkey (TURKEY) Educational Grant, 2021.

Publications from the Last 3 Years:

1. Sonmez A, Haymana C, Bayram F, Salman S, Dizdar OS, Gurkan E, Kargili Carlıoglu A, Barcin C, Sabuncu T, Satman I; TEMD Study Group. Turkish nationwide survEy of glycemic and other Metabolic parameters of patients with Diabetes mellitus (TEMD study). *Diabetes Res Clin Pract*. 2018;146:138-147. doi: 10.1016/j.diabres.2018.09.010.
2. Aslı Gizem Pekmezci, Kürşat Gündoğan, Oğuzhan Sıtkı Dizdar, Emine Alp Meşe. Daily energy and protein intake in hospitalized patients in department of infectious diseases: a prospective observational study. *Progress in Nutrition 2018*; Vol. 20, Supplement 2: 00-00 DOI: 10.23751/pn.v20i2-S.5448.

3. Esmeray K, Dizdar OS, Erdem S, Gunal Aİ. Effect of Strict Volume Control on Renal Progression and Mortality in Non-Dialysis-Dependent Chronic Kidney Disease Patients: A Prospective Interventional Study. *Med Princ Pract.* 2018;27(5):420-427. doi: 10.1159/000493268.
4. Sonmez A, Yumuk V, Haymana C, Demirci I, Barcin C, Kıyıcı S, Güldiken S, Örük G, Ozgen Saydam B, Baldane S, Kutlutürk F, Küçükler FK, Deyneli O, Çetinarslan B, Sabuncu T, Bayram F, Satman I; TEMD Study Group. Impact of Obesity on the Metabolic Control of Type 2 Diabetes: Results of the Turkish Nationwide Survey of Glycemic and Other Metabolic Parameters of Patients with Diabetes Mellitus (TEMD Obesity Study). *Obes Facts.* 2019 Mar 20;12(2):167-178. doi: 10.1159/000496624.
5. Oğuzhan Sitki Dizdar, Ayşe Turunç Özdemir, Osman Başpinar, Derya Koçer, Yavuz Katircilar, İlhami Çelik. Serum prolidase level in patients with brucellosis and its possible relationship with pathogenesis of disease: a prospective observational study. *Turk J Med Sci* (2019) 49: doi: 10.3906/sag-1902-122.
6. Belkıs Nihan Coskun, Oguzhan S. Dizdar, Seniz Korkmaz, Engin Ulukaya, Turkkan Evrensel. The roles of M30 and M65 in the assessment of treatment response and prognosis in patients with non-small cell lung cancer, who receive neoadjuvant treatment. *Contemp Oncol (Pozn)* 2019; 23 (4): 208–213.
7. Dizdar OS, Yeşiltepe A, Dondurmaci E, Ozkan E, Koç A, Gunal AI. Hydration status and blood pressure variability in primary hypertensive patients [published online ahead of print, 2020 Jun 11]. *Nefrologia.* 2020;S0211-6995(20)30035-7. doi:10.1016/j.nefro.2020.02.002.
8. Dizdar OS, Yıldız A, Gul CB, Gunal AI, Ersoy A, Gundogan K. The effect of hemodialysis, peritoneal dialysis and renal transplantation on nutritional status and serum micronutrient levels in patients with end-stage renal disease; Multicenter, 6-month period, longitudinal study. *J Trace Elem Med Biol.* 2020;60:126498. doi:10.1016/j.jtemb.2020.126498.

9. Sonmez A, Tasci I, Demirci I, et al. A Cross-Sectional Study of Overtreatment and Deintensification of Antidiabetic and Antihypertensive Medications in Diabetes Mellitus: The TEMD Overtreatment Study. *Diabetes Ther*. 2020;11(5):1045-1059. doi:10.1007/s13300-020-00779-0.
10. Bakkal H, Dizdar OS, Erdem S, et al. The Relationship Between Hand Grip Strength and Nutritional Status Determined by Malnutrition Inflammation Score and Biochemical Parameters in Hemodialysis Patients [published online ahead of print, 2020 Mar 17]. *J Ren Nutr*. 2020;S1051-2276(20)30029-7. doi:10.1053/j.jrn.2020.01.026.

In: Micronutrients and their Role…
Editor: Horace A. Howard

ISBN: 978-1-53619-843-0
© 2021 Nova Science Publishers, Inc.

Chapter 2

PHYTOCHEMICAL DEFLECTION OF HARMFUL INFLAMMATORY EVENTS TOWARD MORE EFFECTIVE IMMUNE ACTIVITY

Stephen C. Bondy[*], PhD
Department of Occupational and Environmental Health,
Department of Medicine, University of California, Irvine,
Irvine, CA, US

ABSTRACT

A large number of disorders are associated with two changes in immune reactivity. Firstly, a flourishing disease is often accompanied by depression of effective immune surveillance. This shortcoming allows both invasive microorganisms to proliferate and also leads to survival of abnormal cell types, some of which may proliferate and become transformed to malignant variants. A second feature common to several chronic disease states, is the development of ineffective immune

---

[*] Corresponding Author's E-mail: scbondy@uci.edu.

reactions, largely typified by elevation of non-selective generalized inflammatory events. Both of these characteristics adversely affect control and limitation of disease progression. These deficits comprising ever evolving levels of inflammation and a poverty of beneficial immune responses, are also found in normal aging and consequently are especially marked in age-related ailments.

The basis for the use of micronutrient phytochemicals is to optimize health in normal subjects and thus pre-emptively assure that the response to an adverse disease event will be the best possible. The maintenance of a high level of fitness in advance of untoward circumstances, is important in delaying and limiting the consequences of both disease states and changes accompanying senescence.

This report emphasizes the utility of a series of micronutrient phytochemicals that are able to redirect immune responses toward a more specific goal, while diminishing the propensity of inflammatory mechanisms to be over-reactive and inappropriately directed. In general, phytonutrients tend to be less directed toward a single metabolic site than are pharmacological agents, and tend to impact a series of targets. While they are not essential like vitamins and essential minerals, they are likely to have substantial benefits regarding overall human health and longevity as a result of their disease-preventing properties. This breadth of action and generally low toxicity is in contrast to pharmaceuticals which carry a higher risk of adverse side effects. This makes phytochemicals eminently suitable for extended usage over a prolonged period.

INTRODUCTION

Natural products have been used for both health maintenance and for medicinal purposes for several millennia. In the last 150 years they have been increasingly displaced by more purified preparations generally made by procedures common to organic chemistry. These man-made chemicals have the virtue of being free of unwanted contaminants and of generally having greater potency. They can also target a precise molecular event and can thus have a more focused site of action. Pharmacological drugs are normally used in response to the appearance of a distinct disease, the aim being to restore health and well-being by pinpointing the site of impairment and using a selected chemical to target that site. This focal approach is especially useful in cases of an acute injury incurred by an exogenous infection or an endogenous metabolic disturbance with a circumscribed perimeter. New drugs are constantly developed and synthesized, designed to increase the specificity of their actions, and minimize adverse side effects.

The limitations of synthetic drugs are often inherently to be found in the very features that constitute their strengths. Thus, the potency and narrow site of action of such agents can lead to robust metabolic compensatory responses intended to restore cellular homeostasis.

While many drugs are effective in the short term, this tendency of the body to restore equilibrium when metabolism is driven in a novel direction can result in reduced efficacy of action after prolonged application of a drug. The development of such refractory resistance is often combined with the emergence of undesirable side effects. Such effects can be seen in the case of drugs used to treat many chronic conditions such as asthma, Parkinson's disease and schizophrenia, all of which require repeated application of medication. The re-emergence of the underlying condition in time, is often combined with the onset of unwelcome sequelae. Drug resistance is an ongoing problem and is one of the drivers in the continual search for products.

Recently the usefulness of corticosteroids in treating COVID-19 has been brought into question. While reducing mortality on extreme cases, in other situations, this powerful suppressant of immune responses can prolong the survival of the virus in the body (Mudd et al., 2020, Russell et al., 2020). In contrast, the perceived shortcomings of natural products, their lack of specificity, their lower potency, and their often containing a range of related molecular species rather than a single chemical, can also lead to the advent of distinct benefits. All of these features tend to interrupt an undesirable sequence of intracellular changes or promote a beneficial trajectory is a more diffuse manner. A metabolic chain of events is likely impacted on at several sites, which reduces the likelihood of occurrence of a countervailing attempt to reverse effects. Melatonin has been found to be of value in treatment of COVID-19 without appearing to possess the limitations of steroids (Zhou et al., 2019). The generally broader but less intense impact of natural products also allows their use for more extended time periods. Thus, while application of strong pharmacological agents is most suitable for acute situations, whether relating to pain, infection, injury or some form of organ collapse, natural products may be most useful in treatment of chronic disease or in continued maintenance of optimal health through the lifespan.

It is known that the immune system grows less effective with aging and trends toward generation of non-targeted inflammatory changes. This is the major reason for the large increase in mortality in the elderly relative to the young, attributable to COVID-19 infection. This increasing susceptibility to poor immune reactions is a rationale for proposing augmenting the immune response in the aged by the continuing use phytonutrients. This recommendation is not confined to the ailing but also to the health aging population. In contrast to many pharmacological agents, the agents discussed are all well tolerated and suitable for extended use, and have minimal untoward side effects. Furthermore, there is little evidence of the development of metabolic reaction designed to maintain homeostasis by reversing the more restrained effects of these products.

Some general comments apply to nutrient phytochemicals as a whole:

1. Most of the agents to be discussed act by regulation of uncontrolled inflammatory processes and free radical generating oxidant events that typify many chronic disease states and well as senescence (Bondy and Sharman, 2010).
2. It has been difficult to link specific health benefits to specific phytochemicals. The complex nature of these chemicals within plants, and the common heterogeneity of each class of compound, has made epidemiological validation of benefits challenging.
3. Due to the many sites of action often found to be affected by a single phytochemical, pinpointing their precise mechanism of action has also been problematical. Very few studies in the literature compare different phytochemicals so it is difficult to know whether as given agent has a distinctive target of whether many of these agents in fact, act at the same sites.

4. The valuable attributes of these agents are generally not apparent immediately but may take an extended period to become manifest. This further confounds evaluation of their value.
5. Despite all these handicaps to assessment of the utility of phytochemicals, there is broad epidemiological evidence derived from distinct populations, of the health value of a predominantly plant-based diet (Medina-Remón et al., 2018).

Some classes of phytochemical and other biological agents are briefly discussed but rather than primarily stressing the literature on effects of individual compounds, this review addresses the question of those major intracellular processes where phytonutrients might act in a supportive manner. The emphasis of this review is in suggesting that many phytochemicals have a broad resemblance in their key sites of action. If this is the case, more attention needs to be given to their degree of access to intracellular compartments and to the kinetics of their persistence. The sections are thus subdivided so as to center on analysis of the crucial biological events that lead to improved cellular health and effectiveness.

KEY BIOLOGICAL PROCESSES THAT CAN BE MODULATED BY PHYTONUTRIENTS

There are some key networks that have a major impact on the qualitative and quantitative nature of immune responses. These pathways are summarized in Figure 1.

1. *Activity Level of Inflammation-Related Genes.* These include pathways containing transcription factors such as NF-kB involved in activating genes related to both the innate and adaptive immune response systems. This includes enhancement of production of mRNAs for inflammatory cytokines which are widely implicated in many inflammatory diseases.

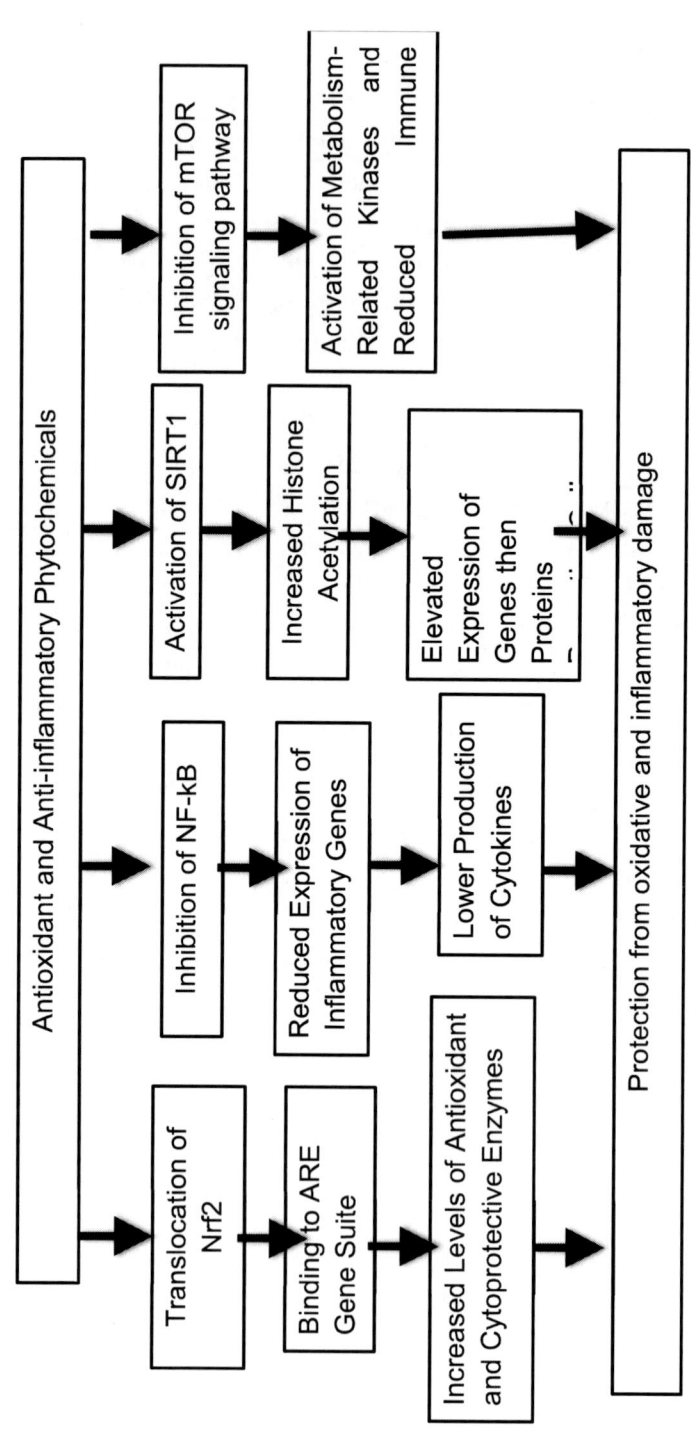

Figure 1. Protective pathways influenced by common phytochemicals.

2. *Expression of antioxidant, anti-inflammatory and detoxification genes.* In contrast other networks support production of antioxidant and anti-inflammatory processes. A premier example of this is the Keap1-Nrf2-antioxidant element response (ARE) pathway. The Keap zinc finger protein binds to the Nrf2 transcription factor and this binding is loosened under condition of oxidative stress, allowing Nrf2 translocation to the nucleus. This allows derepression of DNA sites containing the antioxidant response element and consequent expression of a series of antioxidant and cytoprotective proteins.
3. *Histone deacetylation of sites allowing increased expression of mRNAs supporting cell survival.* Sirtuins are another group of NAD+ activated signaling proteins, several of which can deacetylate nuclear histones and thus weakening their binding to DNA. SIRT1 activity requires NAD+ and enhancing the concentration of this can enhance SIRT1 effectiveness (Xie et al., 2020).

   Sirtuin-1 (SIRT1) levels are elevated in response to various stressors and toxicants including inflammatory and free radical producing events. This enables increased expression of proteins including endocrine factors which promote cell survival. Activation of this suite of genes may be protective against neurodegeneration, vascular inflammation and excessive storage of lipids. However, SIRT1 can have both pro- and anti-inflammatory effects of differing cell types (Chadha et al., 2019).
4. *The mammalian target of rapamycin complex1 (mTORC1) signaling pathway.* This pathway activates a series of kinases and integrates upstream metabolic and energy-related events (LaPlante and Sabatini, 2009). It also down-regulates immune responses. Inhibition of this pathway by rapamycin and its analogs may be beneficial for a range of neurological disorders, cancer and may also slow the onset of age-related changes (Johnson et al., 2013).

TYPES OF PHYTOCHEMICAL WITH PHARMACOLOGICAL POTENTIAL

The compounds are listed, initially with emphasis on their utility to the organisms synthesizing them. Insufficient consideration had been given to the original evolutionary purpose of these chemicals but such information may help to illuminate the mechanisms underlying their potential clinical applicability. Less emphasis is given to the applications of agents for treatment of human disorders as claims are often very broad and generally encompass a large range of diseases. In view of this, the issue of whether compounds act at very similar loci or have the potential to interact in a synergistic manner is relevant and largely unexplored.

Polyphenols

Polyphenols generally promote SIRT1 and Nrf2/ARE activity and inhibit pro-inflammatory events by reducing NF-kB activity (Sarubbo et al., 2018).

Flavonoids such as quercetin consist of a 15-carbon frame consisting of two phenyl rings and a heterocyclic ring. Quercetin is a plant pigment present in onions, grapes, berries, broccoli, and citrus fruits. It can protect plants against oxidative stress in the presence of free-radical generating agents such as paraquat (Zerin et al., 2013). The neuroprotective effects of flavonoids have been attributed to enhancement of Nrf2/ARE and SIRT1 together with inhibition of NF-kB (Dong et al., 2017, Velagapudi et al., 2018).

Resveratrol and Pterostilbene (trans-3,5-dimethoxy-4-hydroxystilbene) are stilbenoid polyphenols, where two phenol rings linked to each other by an ethylene bridge. Pterostilbene and resveratrol are major antioxidant component of grapes, blueberries and peanuts, produced in greater amounts in response to tissue damage and invasive organisms. These agents may predominantly but not exclusively act on the sirtuin class of NAD(+)-dependent deacetylases, (Kane et al.) and are considered a promising new strategy in treatment of many diseases involving oxidant and inflammatory processes, especially those relating to senescence (Sarubbo et al., 2018). One of the beneficial systems by resveratrol is the Nrf2/ARE trajectory (Kim et al., 2018).

Curcumin is a polyphenol made by the *Curcuma longa* plant with antioxidant and anti-inflammatory properties. It is of value to the plant because of its antimicrobial and antifungal properties and its ability to repel various insect species. It is currently being tested in several clinical trials as an adjunct therapy for various forms of cancer (Tomeh et al.,2019). Nrf2/ARE is a key pathway activated by curcumin (Jiang et al., 2020).

Epigallocatechin gallate is a polyphenolic ester found in teas, fruits and dark chocolate. In plants, it is protective against fungi such as *Botrytis cinereal* and can block seed germination under adverse environmental conditions. Both epigallocatechin and resveratrol enhance the Nrf2/ARE track and this may lead to carcinostatic effects in several organs (Smith et al., 2016).

Anthocyanins are brightly pigmented water-soluble diacid flavonoids found in blueberries, raspberries, black rice and many other blue or purple vegetables. They play a role attracting the pollinators and seed dispersers essential for plant reproduction. By virtue of their direct anti-oxidant properties and ability to activate the Nrf2 pathway they are protective against high light stress and their coloration can also serve as a signal to discourage herbivory. Their antioxidant properties maty also account for the attributed value of plants high in anthocyanins in the mitigation of cardiovascular disease (Blesso, 2019).

Isoflavones such as genistein are flavonoid polyphenols found especially in soybeans. Their usefulness to plants is that they induce nodulation in plant-growth promoting rhizobacteria that reside on the roots of the plant (Mabood et al., 2006). These bacteria confer systemic resistance of the plant against the destructive bacterium *Xanthomonas axonopodis pv. Glycines*. As well as stimulation of the Nrf2/ARE pathway (Liang et al., 2018), isoflavones bind to estrogen receptors and possess both weak estrogenic and anti-estrogenic activity. They have low solubility in water and have a bitter taste. The anti-estrogenic properties of isoflavones may reduce the risk of hormone-associated cancers (breast, uterine, and prostate), while estrogenic effects in other tissues may promote maintenance of bone mineral density and lowering of LDL- and raising of HDL- cholesterol. An association has also been suggested between soybean consumption and improved cognition (Cui et al., 2020). On the other hand, low level disruption of estrogenic events may also have undesirable effects (as is shown in the case of bisphenol A) and thus use of this class of agent may thus have negative consequences (Patisaul et al., 2017).

Berberine is a yellow alkaloid polyphenol isoquinoline with a tetracyclic skeleton. It is found in many plants where its value is probably due to its anti-microbial activity. In addition to claims of its utility as an anti-oxidant and anti-inflammatory agent and activator of the Nrf2/ARE system (Ashrafizadeh et al., 2020), it has found more specific use in the treatment of cancer and diabetes. It has been reported as effective as metformin in the treatment of mild Type 2 diabetes. Several reports of berberine toxicity exist (Martini et al., 2020) and thus this agent is potentially more hazardous than most of the others agents discussed.

Organosulfur Compounds

The biological utility of these agents in both the plants that manufacture them and the animals that consume them is thought to be due to redox-reaction with -SH groups in glutathione and proteins, thus preserving intracellular reducing power.

Sulforaphane is an organic isothionate material synthesized in response to injury by cruciferous vegetables including broccoli, cauliflower, Brussels sprouts, cabbage, mustard greens, and watercress. Its promotion of the Nrf2/ARE sequence is a major means of supporting anti-inflammatory events (Calabrese et al., 2020). It has anti-inflammatory and anti-oxidant properties.

Allicin is an organosulfur compound obtained from garlic. It is produced from the amino acid, S-allyl cysteine sulfoxide upon tissue damage to the plant. Allicin can either kill or inhibit proliferation of a range of invasive bacteria and fungi (Müller et al., 2016). Its reducing properties may account for this and for its ability to alter immune cell signaling but allicin has a range of poorly understood properties (Borlinghaus et al., 2014).

Carotenoids

Lycopene is a red carotenoid antioxidant found in some red fruits and vegetables including tomatoes, carrots, and watermelons. Its utility to plants may relate to ultraviolet protection during photosynthesis in sunlight. It has no β-carotene vitamin activity but is able to enhance the Nrf2/ARE pathway and the act as an antioxidant (Inoue et al., 2017a). *Lutein* is a related compound with analogous features.

Ginkgolides

Ginkgolides are diterpenoid lactones found in *Ginkgo biloba*, with 20-carbon structures arranged in six 5-membered rings. These compounds are effective in repelling insect predators from the plant and also inhibit larval detoxifying enzymes including glutathione transferase, acetylcholinesterase, carboxylesterase and mixed function oxidase (Pszczolkowski et al., 2011, Pan et al., 2016). These combined qualities provide the plants with significant protection from insects.

Extracts of *Ginkgo biloba* extract has been used medicinally for over 1000 years for its neuroprotective, anti-cancer, cardioprotective, and stress alleviating properties. In addition to its antioxidant and anti-inflammatory nature enable by activation of the Nrf2 pathway (Liu et al., 2019), its beneficial qualities have been attributed to its specific inhibition of platelet activation factor (PAF) which is involved blood coagulation and the furthering of inflammation. Ginkgo extract and other PAF inhibitors can block the excessive and undesirable immune activity found in allergies and asthma and other types of inflammation (Gerstmeier et al., 2019). Inhibition of PAF has been found to block the aggregation of β-amyloid peptide and ginkgolides have been found useful in the treatment of Alzheimer's disease and cerebrovascular disease (Mango et al., 2016, Liao et al., 2020).

Omega-3 Fatty Acids

These essential polyunsaturated fatty acids are found in many plant and marine oils and characterized by the presence of a double bond three atoms away from their terminal methyl group. Eicosapentaenoic acid (EPA), and docosahexaenoic acid (DHA) are found in fatty fish and other marine organisms while α-linoleic acid is found in a range of nuts and seeds. They are important constituents of cell membranes and also act as activators of several receptors involved in signal transduction. These signaling events tend to reduce inflammatory activity and promote glucose uptake by cells. It is likely that omega-3 fatty acids perform these functions in plants and animals. In addition, as they have a lower freezing point than most lipids, their high content in cold water fish is advantageous in maintaining membrane fluidity at temperatures near freezing. Fish are not able to synthesize omega-3 fatty acids and obtain them from single celled phytoplankton, often by way of the food chain, initially being transferred to zooplankton (De Carvalho et al., 2018).

The use of these lipids has been reported to be beneficial in the treatment of a range of disorders including cardiovascular disease, various forms of dementia and rheumatoid arthritis. Both the SIRT1 and the Nrf2 trajectories are stimulated by these compounds (Inoue et al. 2017b, Zhang et al., 2014).

Melatonin

This is a pleiotropic hormone predominantly released by the pineal in animals, and functions in the diurnal light-dark cycle and in both animals and plants, acting as a potent hormone and immunomodulator (Agathokleous et al., 2019). Melatonin is also made in plants where it can play many roles including regulation of development, seed germination, and fruit ripening. In plants, melatonin can also mount effective immune responses and defenses against many kinds of stressor including drought, extreme temperatures, excessive salinity, and anthropogenic pollution. The widespread occurrence of melatonin among several phyla indicates its early evolutionary origin and its importance. Its oft-reported antioxidant properties are likely to be indirect and mediated by amplifying signaling pathways, as it is not present within cells in sufficient amounts to be a direct anti-oxidant and effective scavenger of free radicals. It is synthesized from tryptophan by both animal and plant species.

Administration of exogenous melatonin has been reported to lead to a broad series of health improvements (Bondy, 2018, Ferlazzo et al., 2020). Rather than inhibiting immune responses, melatonin appears to modulate them in a beneficial direction (Hardeland, 2019, Bondy and Campbell, 2020). A major contention that has been made for melatonin is related to its apparent ability to slow down the normal rate of senescence, most likely at the level of alteration of gene expression (Sharman et al, 2007). A variety of positive outcomes may secondarily emerge from such retardation, such as a reduction of the incidence of age-related disorders including neurodegenerative disease and cancer (Sharman et al., 2011).

THE ALTERATION OF REGULATORY PATHWAYS BY PHYTONUTRIENTS

Phytonutrients can effect substantial regulatory changes in each of the major sequences described above. Table 1 summarizes the many reports of positive effects of various phytochemicals on either inhibition or promotion of these pathways. What is rather remarkable is the degree of overlap between various active agents in their reported sites of action. In addition, each phytonutrient considered individually has considerable breadth of sites of action that are affected.

The antioxidant and anti-inflammatory sequence of events enabled after activation of the Nrf2/ARE pathway is generally helpful in mitigating the effects of many chronic inflammatory states (Ahmed et al., 2017). Similarly promotion of SIRT1 generally leads to improved outcomes. Several phytochemicals that may be helpful in treatment of Alzheimer's disease, act by reducing activity of the NF-κB inflammatory pathway by inhibition of the degradation of IκB and of translocation of NF-κB to the (Seo et al., 2018). However, while suppression of extended NF-kB inflammatory activity is desirable in persistent inflammatory disease, action of this transcription factor is important in integrating an acute inflammatory response to the presence of abnormal cells or exogenous organisms, thus bringing about their elimination.

Table 1. Positive or negative effect of various plant-derived chemicals on key routes by which inflammatory and pro-oxidant events are regulated

Phytochemical	mTOR inhibited	NF-kB inhibited	Nrf2/ARE activated	SIRT-1 activated
Resveratrol/pterostilbene	+and -	+	+	+
Curcumin	+	+	+	+
Epigallocatechin		+	+	+
Quercetin	+	+	+	+
Sulforaphane	-	+	+	+
Allicin	+	+	+	+
Anthocyanins	+	+	0	+
Melatonin	-	+	+	+
Lycopene	+	+	+	+
Ginkgolides	-	+		+
Berberine	+	+	+	0
Isoflavones	+	+	+	+
Omega-3 fatty acids	+	+	+	+

The consequences of inhibition of the mechanistic target of rapamycin (mTOR), a serine/threonine kinase, are equivocal. Its excess content in tissues is associated with several persistent pathological conditions where age is a key factor, including cancer, neurodegenerative disease, arthritis and diabetes. mTORC1, a component of the mTOR complex, drives the expansion of immune T cells and thus promotes inflammation (Perl, 2015). A variety of plant materials depress levels of mTOR and have thus been linked to cancer prevention and improved outcomes of cancer treatment (Badar et al., Naujokat et al., 2020). The powerful immunosuppressant properties of mTOR inhibitors has made then useful in organ transplantation. While their use has been advocated for improving longevity (Blagosklonny, 2019), it should be borne in mind that mTOR signaling has an important integrative function in the regulation of growth, development and cellular energy regulation and its inhibition can lead to toxic consequences (Zhang et al., 2019). Such side-effects are less likely with the phytochemicals listed than with bacterially derived rapamycin which targets mTOR with great specificity.

The mechanisms by which these broad regulatory pathways are affected by phytochemicals is likely to involve several other factors not discussed. These include affecting the expression of microRNAs (McCubrey et al., 2017). For example, treatment with curcumin decreases the expression of miR-21 and miR-34a, while the tumor suppressor let-7a miRNA is upregulated (Mudduluru et al., 2011, Subramanian et al., 2012). Also, modulation of the gut microbiome by phytochemicals is likely to be an important contributor to their health promoting qualities (Blesso, 2019).

LIMITATIONS OF PHYTOCHEMICALS

Bioavailablity

The extent of the bioavailabilty of many of the compounds described is a major problem since they often have very limited solubility in water. Many solutions to this are being studied including the use of nanoparticles, liposomes, emulsions and lipid carriers (Del Prado-Audelo et al., 2019, Ojalide and Sakar, 2020).

Another feature potentially limiting utility is the rate of metabolism of compounds, if this is very rapid, their utility can be limited. The short half-life is intracellular quercetin exemplifies this (Iside et al., 2020) where bioavailability is only around 2% of an oral dose (Costa 2016), while sulforaphane has been reported to be more stable than most phytochemicals (Konrad and Nieman, 2015, Houghton, 2019). Finally, some phytochemicals such as quercetin bind to plasma proteins and this can limit their transfer into cells (Li et al., 2016)

Potential Toxicity of Phytochemicals

The ingestion of large amounts of antioxidant chemicals such as vitamins, lipoic acid and acetyl cysteine can alter redox levels within the cell. This can be beneficial but excessive use of such chemicals can lead to an excess of reducing elements, which nan have adverse effects. (Narasimhan and Rajasekaran, 2015). This imbalance, reductive stress, is unlikely to be caused by most phytochemicals since they do not act directly as antioxidants but rather work by modulation of metabolic signaling pathways (Lee et al., 2014).

The applicability of agents to a specific disorder is not always well understood and this can pose a hazard. For example, activity of the Keap1/Nrf2 pathway leading to derepression of the ARE pathway, is generally beneficial but may be harmful if administered to patients with advanced cancer (Malavolta et al., 2018). Overactivation of this system is associated with drug-resistant cancer, cardiac hypertrophy (Smith et al., 2016). In general, benefits appear to outweigh risks in enabling this pathway, even in many cases of cancer (Qin and Hou, 2016).

Lack of Standardization of Phytochemicals in US Market

The purity and content of phytochemical products are not strongly regulated in the USA, unlike Germany, and this results in great heterogeneity in the composition of commercially available products. Many phytochemicals are marketed as complex and probably inconsistent mixtures.

Large variances have been found in the pharmaceutical quality of ginkgo extracts where differences include outright adulteration, toxic contaminants, and solubility (Kressman et al., 2002). These could contribute to the variable toxicity of ginkgo preparations. Furthermore, phytochemicals like most drugs have a biphasic benefit curve and can be toxic under some circumstances (Mei et al., 2017). Such concerns of purity and undesirable side effects can limit the utility of plant preparations. Epidemiological studies are already subject to a range of confounders and are further handicapped by such issues. Thus, attempted associations of herbal and supplements with cancer risk, involving careful study of large populations (Satia et al., 2009), can yield suggestions of both beneficial and harmful effects but can never be really definitive.

Lack of Detailed Mechanistic and Clinical Research on Natural Products

There is a paucity of valid mechanistic and clinical research that has been conducted on phytochemicals. This may to some degree be attributable to their prevalence; the consequent difficulty of patenting individual products leading to reduced commercial interest. Unfortunately, this leaves the consumer without authoritative scientific advice concerning the utility, or dosage of the host of products available. This is in stark contrast to the detailed instructions and warnings accompanying chemically manufactured pharmaceutical products.

Uniqueness of Each Phytochemical

The effectiveness of each of these compounds has been described as impinging on numerous other targets within the cell. This makes actual number of primary targets of a specific phytochemical very difficult to ascertain. The degree of correspondence between their properties and the distinctive qualities of each agent cannot be determined without more rigorous investigation. A recent example of the complexity of this issue is the emergence of cannabidiol as a phytochemical with potential value for a wide range of ailments including schizophrenia, epilepsy, chronic pain and Parkinson's disease. While a major site of action of CBD is obviously on cannabinoid receptors, around 20 other potential sites of action have been described (Eisenstein, 2019). Thus, a single pure phytochemical is likely to have many targets within the cell.

The selection of endpoints for study differs with each group of investigators. Many claims have been made for the selective value of a given phytochemical product in treatment of a specific disease. However, until phytochemicals are tested by a standardized uniform protocol, true comparisons cannot be made.

CONCLUSION

All four of the pathways selected for this report can under the appropriate circumstances, have both positive or adverse effects on overall wellbeing. Thus NF-kB activity is important in mounting effective immune responses to invasive bacteria and viruses, while SIRT1 has been shown to be either a tumor activator or a tumor suppressor under differing experimental conditions. mTORC1 has an indispensable role in development its continuing activity after the transition from maturation to aging appears to promote development of a range of age-related disorders.

A robust Keap1-Nrf2-ARE signaling system is known to have many positive aspects but its overactivity can lead to multi-drug resistant cancer and cardiac myopathy by inducing 'reductive stress' (Narasimhan and Rajasekaran, 2015). Phytochemicals by virtue of their non-selective targeting of a wide range of sites and their relatively low potency may be less likely to promote the appearance of such adverse events. They are more suitable for extended low level consumption that are more focused and more potent synthetic drugs. These differences also reduce the likelihood of plant derived agents causing major undesirable side effects. However, it is still important to purify such products so that they can be well-defined and characterized, rather than relying on crude botanical extracts with a range of bioactive components.

There seems to be major overlap between the effects of phytonutrients on several key metabolic signaling pathways. In view of this apparent similarity, it may be asked whether a very limited selection of the large range of these phytonutrients can suffice to produce optimal health benefits. The use of several phytochemicals in combination appears to be more effective than a high dose of a single component (Konrad and Neiman, 2015). In future it will be necessary be to define that blend of phytochemicals that together will lead to the most favorable health effects, and thus to create a rational formulation for their optimal admixture. Only with this knowledge can a blend be developed whose components act positively and in a synergistic manner.

Protracted low level therapies need to be found for dealing with the ever-increasing incidence of neurodegenerative disease prevalent in an aging population. The use of non-harsh preparations of plant origin to reduce neuroinflammation is likely to be helpful in this regard (Bondy et al., 2020). In view of their commonality in nature and likelihood of human consumption of several active phytochemical agents for many millennia, it is also more likely that their ingestion is less metabolically disturbing than that of recently synthesized chemicals with are generally designed for significant and targeted disruption of ongoing events.

REFERENCES

Agathokleous, E., Kitao, M., & Calabrese, E. J. (2019). New insights into the role of melatonin in plants and animals. *Chemico-biological interactions*, *299*, 163–167.

Ahmed, S. M., Luo, L., Namani, A., Wang, X. J., & Tang, X. (2017). Nrf2 signaling pathway: Pivotal roles in inflammation. *Biochimica et biophysica acta. Molecular basis of disease*, *1863*(2), 585–597.

Ashrafizadeh, M., Fekri, H. S., Ahmadi, Z., Farkhondeh, T., & Samarghandian, S. (2020). Therapeutic and biological activities of berberine: The involvement of Nrf2 signaling pathway. *Journal of cellular biochemistry*, *121*(2), 1575–1585.

Badar Ul Islam, Khan, M. S., Husain, F. M., Rehman, M. T., Alzughaibi, T. A., Abuzenadah, A. M., Urooj, M., Kamal, M. A., & Tabrez, S. (2020). mTor Targeting by Different Flavonoids for Cancer Prevention. *Current medicinal chemistry*, 10.2174/0929867327666201109122025. Advance online publication.

Blagosklonny M. V. (2019). Rapamycin for longevity: opinion article. Aging, 11(19), 8048–8067.

Blesso C. N. (2019). Dietary Anthocyanins and Human Health. *Nutrients*, *11*(9), 2107.

Bondy, S. C. (2018) Melatonin: Beneficial Aspects and Underlying Mechanisms. In Correia, L., and Meyers, G. (Eds.) *Melatonin: Medical Uses and Role in Health and Disease*. (pp. 277–294). Hauppauge, NY: Nova Press.

Bondy, S. C. & Sharman, E. H. (2010). Melatonin, oxidative stress and the aging brain. In Bondy, S. C. and Maiese, K. (Eds.). *Oxidative Stress in Basic Research and Clinical Practice: Aging and Age-Related Disorders*. (pp. 339-357) Totowa, NJ: Humana Press.

Bondy, S. C., & Campbell, A. (2020). Melatonin and Regulation of Immune Function: Impact on Numerous Diseases. *Current aging science*, *13*(2), 92–101.

Bondy, S. C., Wu, M., & Prasad, K. N. (2020). Attenuation of acute and chronic inflammation using compounds derived from plants.

Experimental biology and medicine (Maywood, N.J.), 153537 0220960690. Advance online publication.

Borlinghaus, J., Albrecht, F., Gruhlke, M. C., Nwachukwu, I. D., & Slusarenko, A. J. (2014). Allicin: chemistry and biological properties. *Molecules (Basel, Switzerland)*, *19*(8), 12591–12618.

Calabrese, E. J., & Kozumbo, W. J. (2020). The phytoprotective agent sulforaphane prevents inflammatory degenerative diseases and age-related pathologies via Nrf2-mediated hormesis. *Pharmacological research*, 105283. Advance online publication.

Chadha, S., Wang, L., Hancock, W. W., & Beier, U. H. (2019). Sirtuin-1 in immunotherapy: A Janus-headed target. *Journal of leukocyte biology*, *106*(2), 337–343.

Costa, L. G., Garrick, J. M., Roquè, P. J., & Pellacani, C. (2016). Mechanisms of Neuroprotection by Quercetin: Counteracting Oxidative Stress and More. *Oxidative medicine and cellular longevity*, *2016*, 2986796.

Cui, C., Birru, R. L., Snitz, B. E., Ihara, M., Kakuta, C., Lopresti, B. J., Aizenstein, H. J., Lopez, O. L., Mathis, C. A., Miyamoto, Y., Kuller, L. H., & Sekikawa, A. (2020). Effects of soy isoflavones on cognitive function: a systematic review and meta-analysis of randomized controlled trials. *Nutrition reviews*, *78*(2), 134–144.

de Carvalho, C., & Caramujo, M. J. (2018). The Various Roles of Fatty Acids. *Molecules (Basel, Switzerland)*, *23*(10), 2583.

Del Prado-Audelo, M. L., Caballero-Florán, I. H., Meza-Toledo, J. A., Mendoza-Muñoz, N., González-Torres, M., Florán, B., Cortés, H., & Leyva-Gómez, G. (2019). Formulations of Curcumin Nanoparticles for Brain Diseases. *Biomolecules*, 9(2), 56.

Dong, F., Wang, S., Wang, Y., Yang, X., Jiang, J., Wu, D., Qu, X., Fan, H., & Yao, R. (2017). Quercetin ameliorates learning and memory via the Nrf2-ARE signaling pathway in d-galactose-induced neurotoxicity in mice. *Biochemical and biophysical research communications*, *491*(3), 636–641.

Eisenstein, M. (2019). The reality behind cannabidiol's medical hype. *Nature 572*, S2-S4.

Ferlazzo, N., Andolina, G., Cannata, A., Costanzo, M. G., Rizzo, V., Currò, M., Ientile, R., & Caccamo, D. (2020). Is Melatonin the Cornucopia of the 21st Century?. *Antioxidants (Basel, Switzerland)*, *9*(11), 1088.

Gerstmeier, J., Seegers, J., Witt, F., Waltenberger, B., Temml, V., Rollinger, J. M., Stuppner, H., Koeberle, A., Schuster, D., & Werz, O. (2019). Ginkgolic Acid is a Multi-Target Inhibitor of Key Enzymes in Pro-Inflammatory Lipid Mediator Biosynthesis. *Frontiers in pharmacology*, *10*, 797.

Hardeland R. (2019). Aging, Melatonin, and the Pro- and Anti-Inflammatory Networks. *International journal of molecular sciences*, *20*(5), 1223.

Houghton C. A. (2019). Sulforaphane: Its "Coming of Age" as a Clinically Relevant Nutraceutical in the Prevention and Treatment of Chronic Disease. *Oxidative medicine and cellular longevity*, *2019*, 2716870.

Inoue, Y., Shimazawa, M., Nagano, R., Kuse, Y., Takahashi, K., Tsuruma, K., Hayashi, M., Ishibashi, T., Maoka, T., & Hara, H. (2017a). Astaxanthin analogs, adonixanthin and lycopene, activate Nrf2 to prevent light-induced photoreceptor degeneration. *Journal of pharmacological sciences*, *134*(3), 147–157.

Inoue, T., Tanaka, M., Masuda, S., Ohue-Kitano, R., Yamakage, H., Muranaka, K., Wada, H., Kusakabe, T., Shimatsu, A., Hasegawa, K., & Satoh-Asahara, N. (2017b). Omega-3 polyunsaturated fatty acids suppress the inflammatory responses of lipopolysaccharide-stimulated mouse microglia by activating SIRT1 pathways. *Biochimica et biophysica acta. Molecular and cell biology of lipids*, *1862*(5), 552–560.

Iside, C., Scafuro, M., Nebbioso, A., & Altucci, L. (2020). SIRT1 Activation by Natural Phytochemicals: An Overview. *Frontiers in pharmacology*, *11*, 1225.

Jiang, C., Luo, P., Li, X., Liu, P., Li, Y., & Xu, J. (2020). Nrf2/ARE is a key pathway for curcumin-mediated protection of TMJ chondrocytes from oxidative stress and inflammation. *Cell stress & chaperones*, *25*(3), 395–406.

Johnson, S. C., Rabinovitch, P. S., & Kaeberlein, M. (2013). mTOR is a key modulator of ageing and age-related disease. *Nature*, *493*(7432), 338–345.

Kane, A. E., & Sinclair, D. A. (2018). Sirtuins and NAD⁺ in the Development and Treatment of Metabolic and Cardiovascular Diseases. *Circulation research*, *123*(7), 868–885.

Kim, E. N., Lim, J. H., Kim, M. Y., Ban, T. H., Jang, I. A., Yoon, H. E., Park, C. W., Chang, Y. S., & Choi, B. S. (2018). Resveratrol, an Nrf2 activator, ameliorates aging-related progressive renal injury. *Aging*, *10*(1), 83–99.

Konrad, M., & Nieman, D. C. (2015). Evaluation of Quercetin as a Countermeasure to Exercise-Induced Physiological Stress. In M. Lamprecht (Ed.), *Antioxidants in Sport Nutrition*. Chapter 10. Boca Raton, FL: CRC Press/Taylor & Francis.

Kressmann, S., Müller, W. E., & Blume, H. H. (2002). Pharmaceutical quality of different Ginkgo biloba brands. *The Journal of pharmacy and pharmacology*, *54*(5), 661–669.

Laplante, M., & Sabatini, D. M. (2009). mTOR signaling at a glance. *Journal of cell science*, *122*(Pt 20), 3589–3594.

Lee, J., Jo, D. G., Park, D., Chung, H. Y., & Mattson, M. P. (2014). Adaptive cellular stress pathways as therapeutic targets of dietary phytochemicals: focus on the nervous system. *Pharmacological reviews*, *66*(3), 815–868.

Li, Y., Yao, J., Han, C., Yang, J., Chaudhry, M. T., Wang, S., Liu, H., & Yin, Y. (2016). Quercetin, Inflammation and Immunity. *Nutrients*, *8*(3), 167.

Liang, F., Cao, W., Huang, Y., Fang, Y., Cheng, Y., Pan, S., & Xu, X. (2019). Isoflavone biochanin A, a novel nuclear factor erythroid 2-related factor 2 (Nrf2)-antioxidant response element activator, protects against oxidative damage in HepG2 cells. *BioFactors (Oxford, England)*, *45*(4), 563–574.

Liao Z, Cheng L, Li X, Zhang M, Wang S, Huo R. Meta-analysis of Ginkgo biloba Preparation for the Treatment of Alzheimer's Disease. *Clin Neuropharmacol*. 2020 Jul/Aug;43(4):93-99.

Liu, Q., Jin, Z., Xu, Z., Yang, H., Li, L., Li, G., Li, F., Gu, S., Zong, S., Zhou, J., Cao, L., Wang, Z., & Xiao, W. (2019). Antioxidant effects of ginkgolides and bilobalide against cerebral ischemia injury by activating the Akt/Nrf2 pathway in vitro and in vivo. *Cell stress & chaperones*, *24*(2), 441–452.

Mabood, F., Souleimanov, A., Khan, W., & Smith, D. L. (2006). Jasmonates induce Nod factor production by Bradyrhizobium japonicum. *Plant physiology and biochemistry: PPB*, *44*(11-12), 759–765.

Malavolta, M., Bracci, M., Santarelli, L., Sayeed, M. A., Pierpaoli, E., Giacconi, R., Costarelli, L., Piacenza, F., Basso, A., Cardelli, M., & Provinciali, M. (2018). Inducers of Senescence, Toxic Compounds, and Senolytics: The Multiple Faces of Nrf2-Activating Phytochemicals in Cancer Adjuvant Therapy. *Mediators of inflammation*, *2018*, 4159013.

Mango, D., Weisz, F., & Nisticò, R. (2016). Ginkgolic Acid Protects against Aβ-Induced Synaptic Dysfunction in the Hippocampus. *Frontiers in pharmacology*, 7, 401.

Martini, D., Pucci, C., Gabellini, C., Pellegrino, M., & Andreazzoli, M. (2020). Exposure to the natural alkaloid Berberine affects cardiovascular system morphogenesis and functionality during zebrafish development. *Scientific reports*, *10*(1), 17358.

McCubrey, J. A., Lertpiriyapong, K., Steelman, L. S., Abrams, S. L., Yang, L. V., Murata, R. M., Rosalen, P. L., Scalisi, A., Neri, L. M., Cocco, L., Ratti, S., Martelli, A. M., Laidler, P., Dulińska-Litewka, J., Rakus, D., Gizak, A., Lombardi, P., Nicoletti, F., Candido, S., Libra, M., Cervello, M. (2017). Effects of resveratrol, curcumin, berberine and other nutraceuticals on aging, cancer development, cancer stem cells and microRNAs. *Aging*, *9*(6), 1477–1536.

Medina-Remón, A., Kirwan, R., Lamuela-Raventós, R. M., & Estruch, R. (2018). Dietary patterns and the risk of obesity, type 2 diabetes mellitus, cardiovascular diseases, asthma, and neurodegenerative diseases. *Critical reviews in food science and nutrition*, *58*(2), 262–296.

Mei, N., Guo, X., Ren, Z., Kobayashi, D., Wada, K., & Guo, L. (2017). Review of Ginkgo biloba-induced toxicity, from experimental studies to human case reports. *Journal of environmental science and health. Part C, Environmental carcinogenesis & ecotoxicology reviews*, *35*(1), 1–28.

Mudd, P. A., Crawford, J. C., Turner, J. S., Souquette, A., Reynolds, D., Bender, D., Bosanquet, J. P., Anand, N. J., Striker, D. A., Martin, R. S., Boon, A., House, S. L., Remy, K. E., Hotchkiss, R. S., Presti, R. M., O'Halloran, J. A., Powderly, W. G., Thomas, P. G., & Ellebedy, A. H. (2020). Distinct inflammatory profiles distinguish COVID-19 from influenza with limited contributions from cytokine storm. *Science advances*, *6*(50), eabe3024.

Mudduluru, G., George-William, J. N., Muppala, S., Asangani, I. A., Kumarswamy, R., Nelson, L. D., & Allgayer, H. (2011). Curcumin regulates miR-21 expression and inhibits invasion and metastasis in colorectal cancer. *Bioscience reports*, *31*(3), 185–197.

Müller, A., Eller, J., Albrecht, F., Prochnow, P., Kuhlmann, K., Bandow, J. E., Slusarenko, A. J., & Leichert, L. I. (2016). Allicin Induces Thiol Stress in Bacteria through S-Allylmercapto Modification of Protein Cysteines. *The Journal of biological chemistry*, *291*(22), 11477–11490.

Narasimhan, M., & Rajasekaran, N. S. (2015). Reductive potential - a savior turns stressor in protein aggregation cardiomyopathy. *Biochimica et biophysica acta*, *1852*(1), 53–60.

Naujokat, C., & McKee, D. L. (2020). The "Big Five" Phytochemicals Targeting Cancer Stem Cells: Curcumin, EGCG, Sulforaphane, Resveratrol and Genistein. *Current medicinal chemistry*, 10.2174/0929867327666200228110738. Advance online publication.

Olajide, O. A., & Sarker, S. D. (2020). Alzheimer's disease: natural products as inhibitors of neuroinflammation. *Inflammopharmacology*, *28*(6), 1439–1455.

Pan, L., Ren, L., Chen, F., Feng, Y., & Luo, Y. (2016). Antifeedant Activity of Ginkgo biloba Secondary Metabolites against Hyphantria cunea Larvae: Mechanisms and Applications. *PloS one*, *11*(5), e0155682.

Patisaul HB. Endocrine disruption by dietary phyto-oestrogens: impact on dimorphic sexual systems and behaviours. *Proc Nutr Soc. 2017 May;76(2)*:130-144.

Perl A. (2016). Activation of mTOR (mechanistic target of rapamycin) in rheumatic diseases. *Nature reviews. Rheumatology*, *12*(3), 169–182.

Pszczolkowski, M. A., Durden, K., Sellars, S., Cowell, B., & Brown, J. J. (2011). Effects of *Ginkgo biloba* constituents on fruit-infesting behavior of codling moth (*Cydia pomonella*) in apples. *Journal of agricultural and food chemistry*, *59*(20), 10879–10886.

Qin, S., & Hou, D. X. (2016). Multiple regulations of Keap1/Nrf2 system by dietary phytochemicals. *Molecular nutrition & food research*, *60*(8), 1731–1755.

Russell, B., Moss, C., Rigg, A., & Van Hemelrijck, M. (2020). COVID-19 and treatment with NSAIDs and corticosteroids: should we be limiting their use in the clinical setting? *Ecancermedicalscience*, *14*, 1023.

Sarubbo F, Esteban S, Miralles A, Moranta D. Effects of Resveratrol and other Polyphenols on Sirt1: Relevance to Brain Function During Aging. *Curr Neuropharmacol.* 2018;16(2):126-136.

Satia, J. A., Littman, A., Slatore, C. G., Galanko, J. A., & White, E. (2009). Associations of herbal and specialty supplements with lung and colorectal cancer risk in the VITamins and Lifestyle study. *Cancer epidemiology, biomarkers & prevention : a publication of the American Association for Cancer Research, cosponsored by the American Society of Preventive Oncology*, *18*(5), 1419–1428.

Seo, E. J., Fischer, N., & Efferth, T. (2018). Phytochemicals as inhibitors of NF-κB for treatment of Alzheimer's disease. *Pharmacological research*, *129*, 262–273.

Sharman, E. H., Bondy, S. C., Sharman, K. G., Lahiri, D., Cotman, C. W., & Perreau, V. M. (2007). Effects of melatonin and age on gene expression in mouse CNS using microarray analysis. *Neurochemistry international*, *50*(2), 336–344.

Sharman, E. H., Sharman, K. G., & Bondy, S. C. (2011). Extended exposure to dietary melatonin reduces tumor number and size in aged male mice. *Experimental gerontology*, 46(1), 18–22.

Smith, R. E., Tran, K., Smith, C. C., McDonald, M., Shejwalkar, P., & Hara, K. (2016). The Role of the Nrf2/ARE Antioxidant System in Preventing Cardiovascular Diseases. *Diseases (Basel, Switzerland)*, 4(4), 34.

Subramaniam, D., Ponnurangam, S., Ramamoorthy, P., Standing, D., Battafarano, R. J., Anant, S., & Sharma, P. (2012). Curcumin induces cell death in esophageal cancer cells through modulating Notch signaling. *PloS one*, 7(2), e30590.

Tomeh, M. A., Hadianamrei, R., & Zhao, X. (2019). A Review of Curcumin and Its Derivatives as Anticancer Agents. *International journal of molecular sciences*, 20(5), 1033.

Velagapudi, R., El-Bakoush, A., & Olajide, O. A. (2018). Activation of Nrf2 Pathway Contributes to Neuroprotection by the Dietary Flavonoid Tiliroside. *Molecular neurobiology*, 55(10), 8103–8123.

Xie, N., Zhang, L., Gao, W., Huang, C., Huber, P. E., Zhou, X., Li, C., Shen, G., & Zou, B. (2020). NAD^+ metabolism: pathophysiologic mechanisms and therapeutic potential. *Signal transduction and targeted therapy*, 5(1), 227.

Zerin, T., Kim, Y. S., Hong, S. Y., & Song, H. Y. (2013). Quercetin reduces oxidative damage induced by paraquat via modulating expression of antioxidant genes in A549 cells. *Journal of applied toxicology: JAT*, 33(12), 1460–1467.

Zhang, M., Wang, S., Mao, L., Leak, R. K., Shi, Y., Zhang, W., Hu, X., Sun, B., Cao, G., Gao, Y., Xu, Y., Chen, J., & Zhang, F. (2014). Omega-3 fatty acids protect the brain against ischemic injury by activating Nrf2 and upregulating heme oxygenase 1. *The Journal of neuroscience: the official journal of the Society for Neuroscience*, 34(5), 1903–1915.

Zhou Y, Hou Y, Shen J, Mehra R, Kallianpur A, Culver DA, Gack MU, Farha S, Zein J, Comhair S, Fiocchi C, Stappenbeck T, Chan T, Eng C, Jung JU, Jehi L, Erzurum S, Cheng F. A network medicine approach to investigation and population-based validation of disease manifestations and drug repurposing for COVID-19. *PLoS Biol.* 2020 Nov 6;*18(11)*:e3000970.

BIOGRAPHICAL SKETCH

Stephen C. Bondy

Affiliation:

Department of Occupational and Environmental Health; Department of Medicine, University of California, Irvine

Education:

BA, Biochemistry, (subsidiaries: Physiology, Zoology, Chemistry), University of Cambridge, England
MS, Biochemistry, University of Birmingham, England
PhD, Biochemistry, University of Birmingham, England
MA, Biochemistry, University of Cambridge, England

Research and Professional Appointments:

1985-Date:	Professor of Toxicology, Program in Toxicology, Center for Occupational and Environmental Health, Department of Medicine, University of California, Irvine, CA
1987-Date:	Professor, Department of Pharmacology, University of California, Irvine, CA
1978-1985:	Associate Professor, Department of Pharmacology, University of North Carolina, Chapel Hill, NC
1978-1985:	Head, Neurochemistry Section, Laboratory of Behavioral and Neurological Toxicology, National Institute of

	Environmental Health Sciences, Research Triangle Park, NC
1976-1978:	Associate Professor of Neurology and Pharmacology, University of Colorado Medical Center, Denver, CO
1970-1976:	Assistant Professor of Neurology, University of Colorado Medical Center, Denver, CO
1965-1970:	Assistant Research Biological Chemist, Department of Biological Chemistry and Associate Member, Brain Research Institute, UCLA School of Medicine, Los Angeles, CA
1963-1965:	Postdoctoral Scientist, Department of Pharmacology, New York State Psychiatric Institute, Columbia University, NY
1959-1962:	Research Assistant, Department of Biochemistry, University of Birmingham, England

Publications from the Last 3 Years:

1. Prasad, K. N. and Bondy, S. C. Dietary fibers and their fermented short-chain fatty acids in prevention of human disease. *Bioactive Carbohydrates and Dietary Fibre* 2019. pii: S0047-6374, 30013-7 doi: 10.1016/j.mad.2018.10.003.
2. Prasad, K. N. and Bondy, S. C. Oxidative and inflammatory events in prion disease: can they be therapeutic targets? Current Aging Sci., 2019, 11:216-225. doi: 10.2174/1874609812666190111100205.
3. Prasad, K. N. and Bondy, S. C. Increased oxidative stress, inflammation, and glutamate: potential preventive and therapeutic targets for hearing disorders. *Mech. Aging. Devel.* 2020, 185:11191. doi: 10.3389/fncel.2017.00276.
4. Bondy, S. C. Aspects of the immune system that impact brain function. *J. Neuroimmunol.* 340, 577167, 2020. doi: 10.1016/j.jneuroim.2020.577167.
5. Nies, I., Hidalgo, K., Bondy, S. C., Campbell, A. Distinctive cellular response to aluminum based adjuvants. *Environ. Toxicol.*

Pharmacol. 78, 103404, 2020. doi.org/10.1016/j.etap.2020.103404.
6. Bondy, S. C. and Campbell, A. Melatonin and regulation of immune function: impact on numerous diseases. *Current Aging Sci.* 2020, 13:92-101. doi: 10.2174/1874609813666200711153223.
7. Bondy, S.C., Wu, M, and Prasad, K. N. Attenuation of acute and chronic inflammation using compounds derived from plants. *J. Exptl. Biol.* 246:406-413, 2021, doi: 10.1177/1535370220960690. Selected as JEB Highlight Article.
8. Bondy, S. C. (ed.) The Molecular Basis for the Environmental Promotion of Neurodegenerative Disease: *Special issue of Int. J. Mol. Sci. 2021.*
9. Bondy, S. and Campbell, A. Aluminum and Alzheimer's Disease. In: *Handbook of Neurotoxicity.* Springer Publishing, NY, (Ed. Kostrzewa, R. M.), 2021.
10. Bondy, S. C. Metal Toxicity and Neuroinflammation In: *Metal Exposure and Neurodevelopmental Disorders.* (Ed. Harischandra, D.), Current Opinion on Toxicol. 2021.

In: Micronutrients and their Role…   ISBN: 978-1-53619-843-0
Editor: Horace A. Howard   © 2021 Nova Science Publishers, Inc.

Chapter 3

MICRONUTRIENTS NEEDS IN CRITICALLY ILL PATIENTS OF PEDIATRIC CARE UNITS

Nazanin Zibanejad[1,2,*], MD

[1]Pediatrics Department, Imam Hossein Children Hospital
[2]Child Growth and Development Research Center, Research Institute for Primordial Prevention of Non-Communicable Disease, Isfahan University of Medical Sciences, Isfahan, Iran

ABSTRACT

Micronutrients are essential nutrient needed by the body in small amounts, including minerals and vitamins. Micro minerals include calcium, phosphorus, magnesium, sodium, potassium, and trace minerals, including magnesium, iron, zinc and selenium. Vitamins and minerals are vital for the prevention of disease, development and health of the body. Only small amounts of them are needed, but they are not produced in the body and must be received through diet.

Deficiency of trace elements such as magnesium zinc phosphate is common in patients admitted to the ICU. Studies have shown that this deficiency is associated with a poor prognosis. Among vitamins for

---

*Corresponding Author's E-mail: zibanejadn@gmail.com.

example vitamin C have antioxidant effects improve inflammation and have a beneficial effect. In acute renal failure, meta-analysis has shown that vitamin C intake reduces the ICU stay and the duration of mechanical ventilation. Low level of Vitamin D have been shown to be strongly associated with problems such as infections, acute liver and kidney damage. However, the effect of vitamin D supplementation is unclear and has not improved the prognosis in one meta-analysis. Zinc levels were significantly lower in patients requiring hospitalization in pediatric ICU than in patients with pneumonia admitted to the ward.

In this section, we will summarize the data of existing articles on the effects of possible benefits and harms of vitamin supplements and essential minerals in pediatric ICU patients and review the results of these studies.

Prescribing micronutrients to pediatric ICU patients can have beneficial effects, including reducing hospital and ICU hospital stays, reducing respiratory illness, and improving infections. In this chapter of the book, we intend to provide an overview of the topic of micronutrients in critically ill children admitted to the PICU.

Keywords: micronutrient, trace element, vitamin, pediatric intensive care unit, critical ill, children

INTRODUCTION

Micronutrients are essential nutrient needed by the body in small amount, including vitamins and minerals [1]. Vitamins and minerals are vital for the prevention of disease, development and health of the body.only small amount of them are needed but they are not produced in the body and must be received through diet.

Micronutrients and macronutrients are essential for metabolism and use by the body and affect almost every enzyme system in the body. Likewise, they are an essential component of a patient's diet and should be taken in the recommended daily allowance [2]. Micronutrients include: zinc, selenium, iron, copper and vitamins: vitamin E, vitamin D, vitamin C, B12, B2, B1 and A[3].

Micronutrients play an important role in mediating metabolism through their function as cofactors in enzymes and as coenzymes, antioxidant systems, and gene transcription.

Decreased serum levels of vitamins during the acute phase response may be due to increased needs (due to increased metabolism as well as due to the use of vitamins for biochemical functions such as protein metabolism), increased catabolism, absorption, increased urinary excretion and potential drug interactions.

After removal of the infection, without any treatment, their level may return to normal serum levels. For example, vitamin A is dependent on retinol-binding protein (RBP) transport. RBP is a negative program and therefore the condition normalizes after the infection clears and thus the vitamin A level improves.

During the acute phase response associated with disease-related inflammation, micronutrients redistribute from the circulatory system under the influence of pro inflammatory cytokines. The decrease in serum concentration is proportional to the concentration of micronutrients should always be accompanied by an index of inflammatory status such as CRP. At present, there is no single universal approach to determining inflammation when determining the level of micronutrients in the body. Inflammation may be classified as minor (CRP< 10 mg/L), moderate (CRP between 11 and 80 mg/L), and major (CRP > 80 mg/L).

Simultaneous determination of the level of reactive protein (CRP) along with micronutrient assessment is necessary to interpret their condition in the patient.

In an acute illness, any deficiency of micronutrient concentrations should also be corrected in an effort to improve clinical outcomes. Correction is necessary due to the effects of micronutrient deficiencies on antioxidant defense mechanisms, metabolic pathways and immune pathways. It is not yet clear whether antioxidant supplements have a clinical advantage in critically ill patients because some studies show clear benefits while others show neutral and even harmful results. However, combination antioxidant therapy makes sense because it works in conjunction with the body's antioxidant network components.

The use of micronutrient supplements in highly ill adults is still controversial. In the pediatric field, due to limited number of studies and confounding factors (e.g., malnutrition or potential risks), the effect of oxidative stress on micronutrient status has been less studied. In order to better understand this phenomenon, in a descriptive study, the status of micronutrients in children with critical illness in terms of nutrition with severe oxidative stress, while the intensity of oxidative stress was increased, there was a significant trend ($p < 0.02$) towards a decrease in plasma micronutrients (selenium, zinc, cooper, vitamin E, vitamin C, and beta-carotene) in severely ill children.

Deficiency or redistribution of multiple micronutrients occurs in very critically ill children with severe oxidative stress. These findings help better identify children who may be taking micronutrient supplements [4].

In those receiving intravenous nutrition, daily intake of multivitamins and micronutrients usually prevents micronutrient deficiencies [5].

Further studies are needed to determine the body's normal severe antioxidant status and the best combination of supplements, either with fortification nutrition or with intravenous and intestinal administration and drug interventions [6].

In this section we summarized the existing data on the effect of possible benefits and harms of vitamin supplement and essential minerals in pediatric intensive care unit (PICU) patients and review the results of these studies.

VITAMIN C

Vitamin C is an antioxidant and a supporter of the immune system and an important cofactor of mono and enzyme deoxygenase. Hypovitaminosis and vitamin C deficiency are very common in critically ill patients due to increased need and decreased intake. Due to several functions of vitamin C and its deficiency can aggravate the severity of the disease and lack of recovery. There is no specific indicator for vitamin C deficiency, but the concentration <0.3 mg/dl. 20 mmol/l indicates an inappropriate and low status of vitamin C in the body [7].

Injectable vitamin C, thiamine, and hydrocortisone have been shown to reduce nosocomial mortality in sepsis patients. However, in a study in the vitamin C group, hydrocortisone and thiamine did not differ much in mortality compared to hydrocortisone administration alone, and the vitamin C administration protocol was not associated with harm to the patient and was associated with a reduction in the period of hospitalization in the intensive care unit; Raises potential profits.

It seems that further studies are needed to evaluate the effect of injecting vitamin C, thiamine and hydrocortisone in patients with sepsis and septic shock as a valuable treatment option [8].

Earlier studies recommended 2 to 3 grams per day of injection (repletion dose) in adults, and organ dysfunction was said to occur less frequently at this dose, but recent studies have suggested 6 to 16 grams per day in adults (pharmacologic dose). Due to this dose, the need for vasopressor and organ dysfunction has been less and has even reduced mortality.

The short course of injectable vitamin C at pharmacological doses seems to be a cheap adjuvant treatment with good tolerance and reliability to regulate severe oxidative stress in severe sepsis, trauma and reperfusion after ischemia [9].

Mild side effects of vitamin C have also been reported. In cancer patients with normal renal function and normal G6PD activity, high-dose vitamin C (one and half grams three times a week) was safe. The dose and duration of vitamin C administration in patients with sepsis is lower and shorter than that prescribed in cancer patients and is safe in these patients.

High doses of vitamin C should be given with caution until the results of further studies in patients with kidney stones, children, G6PD deficient are prepared [10].

A study that has not yet been fully published in children with severe septic shock and was the first study in children suggests that a relatively inexpensive combination of injectable hydrocortisone, ascorbic acid, and thiamine may reduce mortality in children who are accompanied by septic shock [11].

Antioxidant properties, easy and cost-effective use has led to a reluctance to use high doses of vitamin C on patient's beds [12].

In meta-analysis conducted in 2019, 12 trials of 1776 patients showed that vitamin C reduced the ICU hospitalization time by 7.8%. In a number of these studies, oral vitamin C administration of 1 to 3 grams per day reduced ICU admission time by 8.6% and was associated with a reduced duration of mechanical ventilation. Due to the cheapness of vitamin C, this reduction in hospitalization time and mechanical ventilation period is valuable and worth considering [13].

In children, significant benefits of vitamin C supplementation have been observed in the pathology of iron deficiency anemia, depression, and chronic renal failure associated with hemodialysis. No evidence of vitamin C supplementation has been found in morality, cognitive function, infections, and cardiovascular disease. These findings may be due to fact that vitamin C is present in the diet of every person in developed countries, and more studies should done in non-industrialized countries [14].

Although another meta-analysis of 3135 children aged to 3 months to 18 years to evaluate the effect of vitamin C in children in the prevention and treatment of upper respiratory tract infection found no predictive effect, vitamin C intake reduced the duration of upper respiratory tract infections. Due to the frequency of upper respiratory tract infections, improper administration of antibiotics and the safe nature of vitamin C, its supplementation can be justified, especially since these infections and respiratory diseases are one of the main causes of hospitalization in intensive care units, especially in children under 6 years of age and those with frequent upper respiratory tract infections [15].

Injectable vitamin C in adults may reduce the need for vasopressor and mechanical ventilation in critically ill patients without affecting mortality overall.

Hospitalized patients with severe pneumonia who received 6 mg/day of vitamin C protocol had significantly lower hospital mortality than the placebo group and had a much higher radiological improvement on day 7, while vitamin C administration was not associated with increased acute renal failure and superinfection [16].

Early use of intravenous vitamin C in combination with corticosteroids and thiamine is effective in preventing organ dysfunction, including acute kidney damage, reducing mortality in patients with sepsis and septic shock [17].

However, in a meta-analysis in adults, no effect was observed on the rate of infection, length of hospital stay and ICU, duration of mechanical ventilation. However, high doses of vitamin C have shown to reduce mortality. However, the evidence of that meta-analysis does not support the administration of vitamin C supplements in critically ill adult patients [18-19].

However, it seems that sometimes contradictory results in the administration of vitamin C to critically ill patients reported as a whole can be useful in improving the prognosis and reducing mortality in intensive care unit patients, and due to its safe nature, its use is not without benefits.

VITAMIN D

Molecular data suggest that vitamin D should boost internal immunity against bacteria and viruses. Many studies have found a link between vitamin D deficiency and increased infection, autoimmunity and allergies. Studies are also underway on the molecular basis of immune regulation that suggests this association [20].

Many scientific communities have proven that levels above 20 ng/ml25(OH)D of metabolized form of vitamin D are sufficient for bone health, and higher levels of ng/ml 30 are required for optimal extracellular function of vitamin D. About 30% children and 60% of adults in the world are vitamin D deficient and insufficient [21].

A recent meta-analysis suggests an important protective role in vitamin D supplementation in preventing acute viral respiratory infections and exacerbating asthma. In bronchiolitis the results are contradictory and no clear correlation has been found between plasma levels and disease severity. Current data suggest vitamin D supplementation as a cost-effective strategy in reducing infant mortality and morbidity [22].

In a number of adults admitted to the ICU, the relationship between serum vitamin D levels in the first 48 hours of intensive care unit with prognostic markers was investigated. 91.6% of patients were deficient in vitamin D at the time of admission. In this study, no correlation was found between vitamin D level and mechanical ventilation time, prognostic CRP score sofa, APACHE2 and mortality but was associated with charlson comorbidity index (CCI) and vitamin D level was inversely associated with comorbidities such as cancer and liver disease [23].

The protective and therapeutic role of vitamin D has also been widely discussed in the Covid pandemic, and one of these studies, conducted in 2020 on critical Covid 19 patients, found that most critically ill patients had low levels of vitamin C and serum vitamin D. Older age and lower levels of vitamin C were co-dependent risk factors for mortality.

Therefore, in critical ill Covid 19 patients, it is recommended to measure and modify the serum level of vitamin C and D and inject treatment with vitamin C supplements [24].

Inadequate vitamin D levels have been reported in 50% of critically ill patients and are associated with increased mortality, length of hospital stay and ICU, and respiratory involvement with prolonged mechanical ventilation. But in a systematic review involving 695 patients, vitamin D administration did not improve clinical outcomes, so the results are still very contradictory [25].

Recent studies have shown an association between poor vitamin D status >10 ng/ml and respiratory infections and diarrhea in young children and as we know, acute lower respiratory infections (ALRI) and diarrhea are two major causes of mortality in children under 5 years of age.

The risk of ALRI was significantly higher in the vitamin D deficiency group compared to the appropriate vitamin D level group, but during the 6-month follow-up period, vitamin D status was not associated with diarrhea or pneumonia [26-27].

In another systematic review, vitamin D supplementation reduces the risk of acute respiratory infection in all cases. Protective and prophylactic effects are seen in those who receive vitamin D daily or weekly without additional bolus doses but it is not seen in those who receive one or more bolus doses and the general result is that vitamin D supplementation is safe and generally prevents acute respiratory infection. Patients with levels below 25 nmol/l and those who do not receive a bolus dose benefit most from taking a vitamin D supplement.

Although vitamin D deficiency is less than 50 nmol/l in all ARDS patients, it has not been evaluated as a risk factor for ARDS, but the cause of sepsis and its mortality is in the ICU [28].

VITAMIN E

It can be argued that antioxidants are potentially important therapeutic agents in regulating the inflammatory response of ectopic hosts such as SIRS. Antioxidants can be inhibited extracellularly by inhibiting toxic and intracellular ROIs by interrupting lipid peroxidation in membranes as well as by interfering with early inflammatory responses by blocking or modifying the signal transduction of inflammatory cytokines and endotoxins, resulting in cellular activation protect the body.

Systemic inflammatory response syndrome results in destruction of the host tissue and consequent organ failure as a result of uncontrolled overexpression of the host. Oxidative stress as a mechanism for direct cell damage as well as activation of intracellular signaling cascades in inflammatory cells and thus the development of the inflammatory response in this process. In critical patients, a systemic inflammatory process is observed, which can be associated with a decrease in plasma concentrations of antioxidant vitamins. Vitamin E is a cheap, non-toxic and chain antioxidant that has therapeutic potential in regulating this process.

Vitamin E is a general term that includes a mixture of tocopherol and tocotrienol isomers derived from vegetable oils. Tocopherol is the most common from in cell membranes and is biologically active as an antioxidant. Vitamin E has been described as a major antioxidant in mammalian cell membranes that has highly effective antioxidant properties and breaks down membrane lipid peroxidation chains.

Vitamin E is an active oxygen metabolite (ROM). This vitamin is soluble in lipids and its main function is to protect unsaturated fatty acids against oxidative stress.

Overall, studies show that vitamin E is a potent immune modulator in vitro and in vivo with encouraging results in animal models of inflammatory syndromes. Its relative safety, even at high doses, makes it an attractive therapeutic agent [29].

In a study that aimed to evaluate the effect of antioxidant vitamin supplements in critically ill patients and their association with lipid peroxidation, 23 patients received a standard diet (G1) and 11 patients received a daily supplement of 10,000 IU of vitamin A, 400 mg of vitamin E and received 600 mg of vitamin C (G2). The APACHE II score was calculated. Serum concentrations of retinol, beta-carotene, vitamin C and E, malondialdehyde (MDA) and reactive protein C were measured before (T0) and on the eighth day after the start of nutritional therapy (T1). The groups were controlled on T0, T1 and T2 (at discharge or death) on the following parameters: mechanical ventilation, hospitalization days, mortality, incidence of serum MDA and vitamin E infection after intervention and increase in vitamin C in G2 it was significantly less. There was no significant difference between the groups in terms of clinical parameters. As a result, the reported doses of vitamin A, C and E were effective in reducing current lipid peroxidation [30].

In adult patients with mechanical ventilation (ICU) in a prospective randomized clinical trial receiving placebo or 3 IM doses (1000 IU) of vitamin E, vitamin E supplementation appears to be a potent antioxidant, among other measures supportive can be helpful in reducing total SOD activity, ROM production, and risk of organ failure in critically ill patients [31].

ZINC, MAGNESIUM, PHOSPHORUS, SELENIUM

Severely ill patients suffer from severe inflammation and stress in the body, which may increase the use and metabolic replacement of many nutrients, especially zinc. The results of observational studies have shown a worse prognosis for patients with hypomagnesemia, hypophosphatemia and zinc deficiency, but an inverse relationship has been observed between high serum levels of these elements and a worse prognosis [32].

Patients admitted to the intensive care unit often have low magnesium, phosphorus and zinc. Serum levels and supplementation levels are important, but there is no consensus on appropriate levels in children. Measurement and administration of magnesium, zinc, and phosphorus supplements vary significantly among routine ICUs [33].

Plasma levels of critical ill patients are low and associated with organ failure. Because zinc affects inflammation, immune function, and glucose control, zinc supplementation is an acceptable treatment in the intensive care unit [34].

The benefits and harms of magnesium, phosphorus, and zinc supplementation in adult ICU patients are not yet fully understood, and the current systematic review is examining these effects [32].

Serum zinc levels in patients admitted to the ICU for pneumonia were also significantly lower than in patients admitted to the ward. Statistically significant reduction in zinc levels was observed in severely ill children with mechanical ventilation sepsis and deaths. In pediatric pneumonia, serum zinc levels have both diagnostic and prognostic value [35].

Hallmark is sepsis, oxidative stress and dysregulated inflammation. Severely ill patients have lower antioxidant stores, which is associated with increased organ failure and mortality.

Zinc and selenium have important antioxidant function so they can be important in patients with sepsis.

Zinc and selenium are the cornerstones of antioxidant defense in sirs. Serum zinc concentrations are generally low in critically ill children, which is lower in patients with sepsis and multiple organ failure, and zinc levels are inversely related to disease severity. Selenium levels are reduced in patients with sepsis and patients with multiple organ failure [36].

A study was performed to evaluate zinc status in ICU patients with SIRS. Compared to the healthy group in this study, the change in zinc status was associated with the severity and inflammation score in critical ill patients since ICU admission, and SIRS caused a complete cessation of zinc transport in the blood. It was concluded that zinc transporters in the blood may be useful indicators for assessing the severity of systemic inflammation and outcome in critical ill patients [37].

Serum zinc levels in patients with surgical sepsis and surgical cases are also significantly lower than in healthy individuals and this lower level in these patients is more prone to recurrence of sepsis. Patients with surgical sepsis with more organ dysfunction and increased mortality on days 28 and 90 had lower serum levels at the time of admission.

Serum zinc levels are considered acceptable as a biomarker in the diagnosis and evaluation of sepsis patients. However, it is not yet clear whether these findings are duo to overamplified redistribution of zinc during the acute phase response or whether critical ill patients with sepsis also have lower levels of zinc [38].

Oxidative stress due to suboptimal concentrations of selenium and zinc may be associated with damage to key proteins. Zinc deficiency can occur due to low dietary intake, prolonged intravenous feeding without supplements, and intestinal causes such as malabsorption in the ICU. Zinc deficiency is closely linked to stunted growth, respiratory infections, diarrhea and dermatitis [39].

Low zinc levels gave been associated with some degree of organ failure and reduced survival in severely ill children, but basal zinc levels have not been associated with the duration of mechanical ventilation and hospitalization in the intensive care unit and serum zinc had no predictive value for 30-day mortality at the start of ventilation [40].

Zinc deficiency is common in infants with sever pneumonia, but normalization of serum levels with zinc supplementation has not been associated with improved clinical outcome and reduced mechanical ventilation in infants [41].

Taking zinc supplements for more than three months is effective in preventing pneumonia in children under 5 years of age. However, the evidence for its prophylactic effect is not sufficient for a shorter period and also adjuvant treatment with zinc in cases of pneumonia [42].

Following the determination of a safe dose of injectable zinc to normalize serum zinc levels in severely ill children, the 500 mg/kg/daydose of injectable zinc provided approximately 50% of normal plasma zinc levels and was well tolerated [34].

A retrospective study found no association between serum magnesium serum levels and sudden cardiac death, OT interval. However, magnesium above 2.4 mg/dl is a marker predicting an increase in nosocomial mortality in CCU patients [43].

In a systematic review, including 34 studies, the total serum magnesium level has been used to estimate the magnesium status in critical patients. Hypomagnesemia appears to be associated with a higher risk of mortality, but the effect of magnesium supplementation remains unclear and challenging [44].

Selenium plays a key role as a cofactor for glutathione peroxidase and thioredoxin. Selenium deficiency can affect thyroid function and the body's response to oxidative stress. Low levels of selenium are common in critical illness and infection, and a decrease is associated with a worse outcome and increase in sepsis [45-46-47-48-49].

A systematic review study on the effect of injectable selenium in severely ill adult sepsis patients had reduced the risk of mortality from selenium supplementation, but a meta-analysis that updated and included recent RCTs found no difference in mortality with or without selenium [50-51].

There was no difference in the effect of zinc, selenium, glutamine and metoclopramide in children with critical ill, primary outcome including nosocomial infection and clinical sepsis until the first nosocomial infection [52].

Changes in zinc homeostasis and the association between serum zinc levels have been reported in critical patients. The benefits of zinc supplementation have been demonstrated in some types of infectious diseases. However, no evaluate the effect of zinc supplementation in children with sepsis.

An RCT in critical ill children was stopped to evaluate the effect of daily supplementation of zinc, selenium, glutamine, and metoclopramide compared to WHEY protein during administration due to ineffectiveness [52].

Based on conflicting studies in adults, routine zinc supplementation is not recommended in adult ill nutrition guidelines.

Available data on magnesium disorders in critical ill children admitted to pediatric intensive care units, especially in developing countries, are scarce. Hypomagnesemia and hypermagnesemia were observed in 60% and 4% of PICU patients, respectively. The incidence of low RBC magnesium was 3.17 per days. Hypo-magnesium was common in cases of intracranial pressure, and mortality was 9 times higher in patients with hypo-magnesium than in normal magnesium. If both calcium and magnesium were low, the mortality rate was 33% compared to normal levels in both cases. Therefore, hypo-magnesium and RBC magnesium levels are common in PICU patients and are associated with increased mortality [53].

CONCLUSION

Micronutrients and vitamins are essential for metabolism and use by the body and affect almost every enzyme system in the body. They are an essential components of a patient's diet and should be taken as recommended daily. Micronutrients play an important role in mediating metabolism through their function as cofactors in enzymes and as coenzymes in antioxidant systems and gene transcription.

Decreased serum levels of vitamins during the acute phase response may be due to increased needs (due to increased metabolism as well as due to the use of vitamins for biochemical functions such as protein metabolism), increased catabolism, absorption, increased urinary excretion and potential drug interactions.

They may return to normal serum levels after infection clears without any treatment. However, given the review of current articles, it appears that due to the safe nature of micronutrients and vitamins in nutritional and even therapeutic periods due to the many benefits mentioned in most articles for the course of treatment and length of hospital stay, their use can be useful and their prescription is recommended for the intensive care unit.

REFERENCES

[1] https://www.sciencedaily.com/terms/micronutrient.htm
[2] https://www.cdc.gov/nutrition/micronutrient-malnutrition/micronutrients/index.html
[3] https://www.health.harvard.edu/staying-healthy/micronutrients-have-major impact-on-health
[4] Valla FV, Bost M. Multiple Micronutrient Plasma Level Changes Are Related to Oxidative Stress Intensity in Critically Ill Children. *PediatrCrit Care Med.* 2018 Sep; 19(9):e455-e463. doi: 10.1097/PCC.0000000000001626. PMID: 29923936.
[5] Blaauw R, Osland E. Parenteral Provision of Micronutrients to Adult Patients: An Expert Consensus Paper. *JPEN J Parenter Enteral Nutr.* 2019 Mar; 43Suppl 1:S5-S23. doi: 10.1002/jpen.1525. PMID: 30812055.
[6] Koekkoek WA, vanZanten AR. Antioxidant Vitamins and Trace Elements in Critical Illness. *NutrClinPract.* 2016 Aug; 31(4):457-74. doi: 10.1177/0884533616653832. Epub 2016 Jun 16. PMID: 27312081.
[7] Berger MM. Vitamin C requirements in parenteral nutrition. *Gastroenterology.* 2009 Nov; 137(5 Suppl):S70-8. doi: 10.1053/j.gastro.2009.08.012. PMID: 19874953.
[8] Mitchell AB, Ryan TE. Vitamin C and Thiamine for Sepsis and Septic Shock. *Am J Med.* 2020 May; 133 (5):635-638. doi: 10.1016/j.amjmed.2019.07.054. Epub 2019 Aug 28. PMID: 31469984.

[9] Man An, Elbers Pa. Vitamin C: Should we supplement. *Current Opinion in Critical Care.*2018 Aug;24.1.10.1097/MCC.00000000 00000510.

[10] Khoshnam-Rad N, Khalili H. Safety of vitamin C in sepsis: a neglected topic. *Curr Opin Crit Care.* 2019 Aug; 25(4):329-333. doi: 10.1097/MCC.0000000000000622. PMID: 31107310.

[11] Wald EL, Sanchez-Pinto LN. Hydrocortisone-Ascorbic Acid-Thiamine Use Associated with Lower Mortality in Pediatric Septic Shock. *Am J Respir Crit Care Med.* 2020 Apr 1; 201(7):863867. doi: 10.1164/rccm.201908-1543LE. PMID: 31916841.

[12] Collie JTB, Greaves RF. Vitamin C measurement in critical illness: challenges, methodologies and quality improvements. *Clin Chem Lab Med.* 2020 Mar 26; 58(4):460-470. doi: 10.1515/cclm2019-0912. PMID: 31829967.

[13] Hemilä H, Chalker E. Vitamin C Can Shorten the Length of Stay in the ICU: A Meta-Analysis. *Nutrients.* 2019 Mar 27; 11(4):708. doi: 10.3390/nu11040708. PMID: 30934660; PMCID: PMC6521194.

[14] Pecoraro L, Martini L. Vitamin C: should daily administration keep the paediatrician away? *Int J Food SciNutr.* 2019 Jun; 70(4):513-517. doi: 10.1080/09637486.2018.1540557. Epub 2018 Dec 4. PMID: 30513006.

[15] Vorilhon P, Arpajou B. Efficacy of vitamin C for the prevention and treatment of upper respiratory tract infection. A meta-analysis in children. *Eur J Clin Pharmacol.* 2019 Mar; 75(3):303311. doi: 10. 1007/s00228-018-2601-7. Epub 2018 Nov 21. PMID: 30465062.

[16] Langlois PL, Manzanares W. Vitamin C Administration to the Critically Ill: A Systematic Review and Meta-Analysis. *JPEN J Parenter Enteral Nutr.* 2019 Mar; 43(3):335-346. doi: 10.1002/jpen. 1471. Epub 2018 Nov 19. PMID: 30452091.

[17] Marik PE, Khangoora V. Hydrocortisone, Vitamin C, and Thiamine for the Treatment of Severe Sepsis and Septic Shock: A Retrospective Before-After Study. *Chest.* 2017 Jun; 151(6):1229-1238. doi: 10.1016/j.chest.2016.11.036. Epub 2016 Dec 6. PMID: 27940189.

[18] Zhang M, Jativa DF. Vitamin C supplementation in the critically ill: A systematic review and meta-analysis. *SAGE Open Med.* 2018 Oct 19; 6:2050312118807615. doi: 10.1177/2050312118807615. PMID: 30364374; PMCID: PMC6196621.

[19] Kim WY, Jo EJ. Combined vitamin C, hydrocortisone, and thiamine therapy for patients with severe pneumonia who were admitted to the intensive care unit: Propensity score-based analysis of a before-after cohort study. *J Crit Care.* 2018 Oct; 47:211-218. doi: 10.1016/j.jcrc.2018.07.004. Epub 2018 Jul 5. PMID: 30029205.

[20] Mailhot G, White JH. Vitamin D and Immunity in Infants and Children. *Nutrients.* 2020 Apr 27; 12(5):1233. doi: 10.3390/nu12051233. PMID: 32349265; PMCID: PMC7282029.

[21] Peroni DG, Trambusti I. Vitamin D in pediatric health and disease. *Pediatr Allergy Immunol.* 2020 Feb; 31Suppl 24:54-57. doi: 10.1111/pai.13154. PMID: 32017212.

[22] Cepeda S J, Zenteno A D. Vitamin D and pediatrics respiratory diseases. *Rev ChilPediatr.* 2019; 90(1):94-101. Spanish. doi: 10.32641/rchped.v90i1.747. PMID: 31095224.

[23] Gomes TL, Fernandes RC. Low vitamin D at ICU admission is associated with cancer, infections, acute respiratory insufficiency, and liver failure. *Nutrition.* 2019 Apr; 60:235-240. doi: 10.1016/j.nut.2018.10.018. Epub 2018 Oct 24. PMID: 30682545.

[24] Arvinte C, Singh M, Marik PE. Serum Levels of Vitamin C and Vitamin D in a Cohort of Critically Ill COVID-19 Patients of a North American Community Hospital Intensive Care Unit in May 2020: A Pilot Study. *Med Drug Discov.* 2020; 8:100064. doi:10.1016/j.medidd.2020.100064

[25] Langlois PL, SzwecC.Vitamin D supplementation in the critically ill: A systematic review and meta-analysis. *Clin Nutr.* 2018 Aug; 37(4):1238-1246. doi: 10.1016/j.clnu.2017.05.006. Epub 2017 May 11. PMID: 28549527.

[26] Chowdhury R, Taneja S. Vitamin-D deficiency predicts infections in young north Indian children: A secondary data analysis. *PLoS One.*

2017 Mar 8; 12(3):e0170509. doi: 10.1371/journal.pone.0170509. PMID: 28273084; PMCID: PMC5342185.

[27] Martineau AR, Jolliffe DA. Vitamin D supplementation to prevent acute respiratory tract infections: systematic review and meta-analysis of individual participant data. *BMJ.* 2017 Feb 15; 356: i6583. doi: 10.1136/bmj.i6583. PMID: 28202713; PMCID: PMC5310969.

[28] Dancer RC, Parekh D. Vitamin D deficiency contributes directly to the acute respiratory distress syndrome (ARDS). *Thorax.* 2015 Jul; 70(7):617-24. doi: 10.1136/thoraxjnl-2014-206680. Epub 2015 Apr 22. PMID: 25903964; PMCID: PMC4484044.

[29] Bulger EM, Maier RV. An argument for Vitamin E supplementation in the management of systemic inflammatory response syndrome. *Shock.* 2003 Feb; 19(2):99-103. doi: 10.1097/00024382-200302000-00001. PMID: 12578114.

[30] Nogueira CR, Borges F. Effects of supplementation of antioxidant vitamins and lipid peroxidation in critically ill patients. *Nutr Hosp.* 2013 Sep-Oct; 28(5):1666-72. doi: 10.3305/nh.2013.28.5.6590. PMID: 24160231.

[31] Ziaie S, Jamaati H. The Relationship Between Vitamin E Plasma and BAL Concentrations, SOD Activity and Ventilatory Support Measures in Critically Ill Patients. *Iran J Pharm Res.* 2011 Fall; 10(4):953-60. PMID: 24250434; PMCID: PMC3813061.

[32] 32. Vesterlund GK, Thomsen T. Effects of magnesium, phosphate and zinc supplementation in ICU patients-Protocol for a systematic review. *Acta Anaesthesiol Scand.* 2020 Jan; 64(1):131-136. doi: 10.1111/aas.13468. Epub 2019 Oct 13. PMID: 31506930.

[33] Vesterlund GK, Ostermann M. Preferences for the measurement and supplementation of magnesium, phosphate and zinc in ICUs: The international Why Trace survey. *Acta Anaesthesiol Scand.* 2021 Mar; 65(3):390-396. doi: 10.1111/aas.13738. Epub 2020 Nov 25. PMID: 33165935.

[34] Cvijanovich NZ, King JC.Safety and Dose Escalation Study of Intravenous Zinc Supplementation in Pediatric Critical Illness. *JPENJ Parenter Enteral Nutr.* 2016 Aug; 40(6):860-8. doi: 10.1177/ 0148607115572193. Epub 2015 Feb 19. PMID: 25700179; PMCID: PMC5609528.

[35] Saleh NY, Abo El Fotoh WMM. Low serum zinc level: The relationship with severe pneumonia and survival in critically ill children. *Int J Clin Pract.* 2018 Jun; 72(6):e13211. doi: 10.1111/ijcp.13211. Epub 2018 May 31. PMID: 29855123.

[36] Negm FF, Soliman DR. Assessment of serum zinc, selenium, and prolactin concentrations in critically ill children. *Pediatric Health Med Ther.* 2016 Apr 4; 7: 17-23. doi: 10.2147/PHMT.S99191. PMID: 29388624; PMCID: PMC5683293.

[37] Florea D, Molina-López J. Changes in zinc status and zinc transporters expression in whole blood of patients with Systemic Inflammatory Response Syndrome (SIRS). *J Trace Elem Med Biol.* 2018 Sep; 49:202-209. doi: 10.1016/j.jtemb.2017.11.013. Epub 2017 Nov 26. PMID: 29199035.

[38] Hoeger J, Simon TP. Persistent low serum zinc is associated with recurrent sepsis in critically ill patients - A pilot study. *PLoS One.* 2017 May 4; 12(5):e0176069. doi: 10.1371/journal.pone.0176069. PMID: 28472045; PMCID: PMC5417428.

[39] Willoughby JL, Bowen CN. Zinc deficiency and toxicity in pediatric practice. *Curr Opin Pediatr.* 2014 Oct; 26(5):579-84. doi: 10.1097/MOP.0000000000000132. PMID: 25029226.

[40] Linko R, Karlsson S, Pettilä V. Serum zinc in critically ill adult patients with acute respiratory failure. *Acta Anaesthesiol Scand.* 2011 May; 55(5):615-21. doi: 10.1111/j.1399-6576.2011.02425.x. PMID: 21827444.

[41] Yuan X, Qian SY, Li Z, Zhang ZZ. Effect of zinc supplementation on infants with severe pneumonia. *World J Pediatr.* 2016 May; 12(2):166-9. doi: 10.1007/s12519-015-0072-9. Epub 2015 Dec 18. PMID: 26684319.

[42] Sakulchit T, Goldman RD. Zinc supplementation for pediatric pneumonia. *Can Fam Physician.* 2017 Oct; 63(10):763-765. PMID: 29025801; PMCID: PMC5638472.

[43] Naksuk N, Hu T, Krittanawong C. Association of Serum Magnesium on Mortality in Patients Admitted to the Intensive Cardiac Care Unit. *Am J Med.* 2017 Feb; 130(2):229.e5-229.e13. doi: 10.1016/j.amjmed.2016.08.033. Epub 2016 Sep 14. PMID: 27639872.

[44] Fairley J, Glassford NJ, Zhang L, Bellomo R. Magnesium status and magnesium therapy in critically ill patients: A systematic review. *J Crit Care.* 2015 Dec; 30(6):1349-58. doi: 10.1016/j.jcrc.2015.07.029. Epub 2015 Jul 31. PMID: 26337558.

[45] Mertens K, Lowes DA. Low zinc and selenium concentrations in sepsis are associated with oxidative damage and inflammation. *Br J Anaesth.* 2015 Jun; 114(6):990-9. doi: 10.1093/bja/aev073. Epub 2015 Mar 31. PMID: 25833826.

[46] Stadtman TC. Selenocysteine. *Annu Rev Biochem.* 1996;65:83–100.

[47] Iglesias SB, Leite HP, Paes AT et al. Low plasma selenium concentrations in critically ill children: the interaction effect between inflammation and selenium deficiency. *Crit Care.* 2014;18:R101.

[48] Loui A, Raab A, Braetter P et al. Selenium status in term and preterm infants during the first months of life. *Eur J ClinNutr.* 2008; 62:349–355.

[49] Sammalkorpi K, Valtonen V, Alfthan G et al. Serum selenium in acute infections. *Infection.* 1988; 16:222–224

[50] Scott W, Mark J. Surviving sepsis campaign international guidelines for the management of septic shock and sepsis-associated organ dysfunction in children. *Intensive care med.* 2020;46(Suppl 1):S10-S67

[51] Alhazzani W, Almasoud A, Jaeschke R et al. Small bowel feeding and risk of pneumonia in adult critically ill patients: a systematic review and meta-analysis of randomized trials. *Crit Care.* 2013; 17:R127.

[52] Carcillo JA, Dean JM. National Institute of Child Health and Human Development (NICHD). Collaborative Pediatric Critical Care Research Network (CPCCRN) et al. The randomized comparative pediatric critical illness stress-induced immune suppression (CRISIS) prevention trial. *Pediatr Crit Care Med.* 2012; 13:165–173.

[53] Singhi SC, Singh J, Prasad R. Hypo- and hypermagnesemia in an Indian Pediatric Intensive Care Unit. *J Trop Pediatr.* 2003 Apr; 49(2):99-103. doi: 10.1093/tropej/49.2.99. PMID: 12729292.

INDEX

A

acid, 9, 13, 18, 19, 20, 22, 23, 35, 36, 62, 66
acute renal failure, ix, 23, 82, 87
acute respiratory distress syndrome, 98
anemia, 3, 4, 9, 10, 14, 28, 86
antioxidant, ix, 5, 6, 17, 21, 22, 28, 57, 58, 59, 61, 63, 64, 66, 73, 77, 82, 83, 84, 89, 90, 91, 94, 98
arsenic, 4, 15, 16, 21, 39
ascorbic acid, 5, 18, 22, 23, 85
atherosclerosis, 4, 6, 13, 16, 30

B

bacteria, 59, 61, 68, 87
basal ganglia, 15
beneficial effect, ix, x, 21, 82
benefits, vii, ix, 52, 53, 54, 66, 69, 82, 83, 84, 86, 87, 91, 93, 95
bioavailability, 27, 32, 66
biochemistry, 70, 74
biological activities, 70
biological processes, 6
biological samples, 16
biomarkers, 7, 31, 76
blood, 7, 11, 14, 15, 16, 17, 21, 37, 39, 48, 61, 92, 99
body mass index, 10
body weight, 16, 23, 34
bone, 3, 4, 60, 87

C

cadmium, 14, 15, 16, 21
calcium, ix, 9, 15, 81, 94
cancer, 3, 20, 57, 59, 60, 63, 65, 66, 67, 69, 74, 85, 88, 97
cardiovascular disease, 2, 3, 9, 11, 12, 13, 19, 26, 29, 32, 35, 59, 62, 74, 86
cardiovascular risk, 12, 13
cardiovascular system, 74
carotenoid, 61
catabolism, 10, 11, 83, 94
cell biology, 72
cell death, 77
cell membranes, 62, 89
cell metabolism, 15, 30
cell signaling, 61

Index

cellular energy, 65
cellular homeostasis, 53
cellular immunity, 20
central nervous system, 16
cerebrovascular disease, 3, 62
chemicals, 52, 54, 58, 64, 66, 69
children, x, 31, 81, 82, 84, 85, 86, 87, 88, 91, 92, 93, 94, 95, 96, 97, 99, 100
chronic kidney disease (CKD), v, vii, 1, 2, 4, 5, 6, 7, 8, 11, 13, 16, 21, 22, 23, 24, 25, 26, 27, 28, 30, 31, 32, 35, 36, 39, 44, 45, 46, 48
chronic kidney failure, 23
chronic renal failure, 29, 33, 86
cobalamin, 5, 13, 22, 23
cognitive function, 3, 10, 19, 34, 71, 86
cognitive impairment, 19
cognitive performance, 6
compounds, 55, 58, 61, 62, 65, 66, 68, 70, 80
continuous renal replacement therapy, 2, 8, 18, 40
controlled trials, 7, 34, 71
controversial, viii, 2, 4, 13, 84
copper, 3, 4, 5, 6, 12, 14, 16, 17, 18, 21, 24, 28, 40, 43, 82
COVID-19, 53, 54, 75, 76, 78, 97
critical ill, 24, 82, 84, 88, 91, 92, 93, 94, 95, 96, 99, 101
cytokines, 20, 55, 83, 89

D

deficiency, ix, 2, 3, 4, 6, 8, 10, 12, 13, 16, 17, 18, 19, 20, 21, 24, 25, 27, 29, 30, 31, 33, 36, 40, 41, 42, 81, 83, 84, 86, 91, 92, 93, 97, 98, 99, 100
diabetes, 8, 15, 21, 28, 29, 46, 60, 65
dialysis, viii, 2, 4, 5, 6, 7, 8, 9, 10, 11, 13, 15, 16, 17, 22, 24, 26, 28, 29, 32, 33, 35, 36, 37, 48

diet, vii, ix, 4, 9, 10, 12, 15, 19, 23, 25, 26, 55, 81, 82, 86, 90, 94
dietary intake, 3, 4, 9, 10, 12, 13, 19, 41, 92
dietary supplementation, 39
disease progression, viii, 52
diseases, 13, 23, 29, 47, 58, 59, 71, 80, 86, 94, 97
drugs, 16, 52, 53, 67, 69

E

encephalopathy, 3, 4, 14
endothelial dysfunction, 13, 16
end-stage renal disease, 8, 28, 32, 44, 48
energy, 9, 12, 16, 24, 47, 57
evidence, 7, 8, 13, 16, 19, 24, 25, 26, 44, 54, 55, 86, 87, 93
experimental condition, 68

F

fatty acids, 62, 64, 77, 79, 90
folate, 7, 13, 19, 22, 23, 24
folic acid, 5, 9, 13, 18, 19, 22, 23, 35
food, vii, 1, 9, 10, 16, 62, 74, 76
food chain, 62
food intake, vii, 1, 9, 10, 16

G

gene expression, 13, 63, 76
Ginkgo, 61, 73, 75, 76
glutathione, 5, 16, 38, 60, 61, 93
guidelines, 23, 26, 32, 44, 45, 94, 100

H

health, vii, viii, ix, 2, 29, 52, 54, 55, 63, 65, 69, 75, 81, 82, 87, 95, 97

hemodialysis, 2, 3, 5, 12, 14, 28, 31, 32, 33, 34, 35, 36, 37, 38, 39, 40, 44, 48, 49, 86
homeostasis, 6, 7, 29, 30, 54, 93
homocysteine, 13, 19, 22, 32, 41
hormone, viii, 2, 5, 60, 63
hospitalization, ix, 82, 85, 86, 90, 92
human, ix, 27, 28, 29, 32, 52, 58, 69, 75, 79

I

ICU, vii, ix, x, 81, 82, 86, 87, 88, 89, 90, 91, 92, 96, 97, 98
immune function, 6, 80, 91
immune reaction, viii, 52, 54
immune regulation, 87
immune response, vii, viii, ix, 20, 30, 42, 52, 53, 54, 55, 57, 63, 68
incidence, 4, 17, 63, 69, 90, 94
infection, 3, 17, 52, 54, 83, 86, 87, 88, 90, 93, 95, 96
inflammation, vii, viii, ix, 1, 2, 4, 9, 10, 16, 19, 21, 22, 27, 32, 34, 42, 49, 52, 55, 57, 62, 65, 70, 72, 73, 74, 79, 80, 82, 83, 91, 92, 100
inflammatory cells, 89
inflammatory disease, 55, 64
inflammatory mediators, 5
inhibition, 3, 58, 61, 64, 65
injury, iv, 3, 52, 54, 60, 73, 74, 77
intensive care unit, 18, 82, 84, 85, 86, 87, 88, 91, 92, 94, 95, 97
iron, ix, 4, 5, 6, 9, 12, 28, 35, 81, 82, 86

K

kidney, ix, 2, 3, 7, 8, 15, 16, 18, 19, 20, 21, 24, 25, 26, 27, 31, 40, 41, 42, 43, 82, 85, 87
kidney failure, 21
kidney stones, 85

kidney transplantation, 2, 8, 15, 19, 20, 21, 25

L

lipid peroxidation, 5, 7, 31, 89, 90, 98
lipids, 57, 62, 72, 90
longevity, ix, 52, 65, 70, 71, 72
longitudinal study, 28, 48
low-grade inflammation, 10

M

magnesium, ix, 5, 81, 91, 93, 94, 98, 100
magnetic resonance, 15
magnetic resonance image, 15
malnutrition, vii, 1, 2, 8, 9, 11, 16, 22, 26, 32, 38, 84, 95
management, viii, 2, 23, 32, 37, 44, 98, 100
manganese, 4, 5, 7, 14, 16, 31, 37
measurement, 96, 98
mechanical ventilation, ix, 82, 86, 87, 88, 90, 91, 92
mental status change, 3
meta-analysis, ix, 14, 16, 19, 34, 35, 37, 71, 82, 86, 87, 93, 96, 97, 98, 100
metabolic pathways, 83
metabolism, vii, viii, 1, 2, 4, 5, 9, 21, 30, 33, 42, 53, 66, 77, 82, 83, 94
metalloenzymes, viii, 2, 5, 6, 15
micronutrient, vii, viii, 1, 2, 3, 4, 5, 6, 7, 8, 9, 11, 12, 18, 22, 25, 28, 32, 40, 48, 52, 82, 83, 84, 95
minerals, vii, ix, 2, 5, 12, 21, 25, 27, 39, 52, 81, 82, 84
morbidity, vii, 1, 2, 8, 9, 14, 88
morphogenesis, 74
mortality, vii, 1, 2, 4, 7, 8, 9, 13, 14, 16, 17, 19, 20, 22, 26, 32, 42, 43, 53, 54, 85, 86, 87, 88, 89, 90, 91, 92, 93, 94
mortality rate, 94

N

Nrf2, 57, 58, 59, 60, 61, 62, 64, 66, 69, 70, 71, 72, 73, 74, 76, 77
nutrient, ix, 7, 12, 16, 30, 33, 54, 81, 82
nutrition, 4, 19, 23, 24, 27, 32, 40, 44, 45, 46, 74, 76, 84, 94, 95
nutritional status, 10, 16, 25, 28, 38, 48

O

organ, 54, 65, 85, 87, 89, 90, 91, 92, 100
organosulfur, 60, 61
outcomes, vii, viii, 2, 4, 7, 13, 17, 19, 23, 26, 41, 43, 63, 64, 65, 83, 88
oxidative damage, 6, 31, 73, 77, 100
oxidative stress, viii, 2, 3, 4, 5, 6, 9, 13, 15, 16, 21, 22, 27, 29, 57, 58, 70, 72, 79, 84, 85, 90, 91, 93

P

pain, 3, 54, 68
peripheral blood, 20
peripheral blood mononuclear cell, 20
peripheral neuropathy, 3, 14, 17
peripheral vascular disease, 3
peritoneal dialysis, 2, 3, 5, 9, 28, 32, 33, 35, 37, 48
pharmacological agents, ix, 52, 53, 54
plants, 54, 58, 59, 60, 61, 62, 63, 70, 80
pneumonia, ix, 82, 86, 88, 91, 92, 93, 97, 99, 100
polyphenols, 9, 58, 59, 76
polyunsaturated fat, 62, 72
polyunsaturated fatty acids, 62, 72
population, 4, 7, 8, 10, 14, 17, 24, 25, 27, 28, 78
prevention, ix, 24, 65, 76, 79, 81, 82, 86, 96, 101

prognosis, ix, 48, 81, 87, 91
pro-inflammatory, 10, 30, 58
pyridoxine, 5, 13, 18, 22, 23

R

rapamycin, 57, 65, 70, 76
receptors, 59, 62, 68
renal dysfunction, 18
renal failure, 3, 23, 24, 33, 45
renal replacement therapy (RRT), vii, 1, 2, 4, 5, 8, 11, 18, 19, 23, 24, 25, 44
response, viii, 15, 16, 20, 34, 37, 48, 52, 57, 58, 60, 64, 73, 79, 83, 89, 92, 93, 94, 98
risk, vii, ix, 1, 2, 3, 4, 7, 8, 12, 14, 16, 17, 20, 21, 22, 25, 26, 31, 32, 35, 36, 38, 52, 60, 67, 74, 76, 88, 89, 90, 93, 100

S

selenium, ix, 3, 4, 5, 6, 7, 8, 11, 12, 14, 16, 17, 21, 23, 24, 27, 28, 32, 38, 39, 43, 81, 82, 84, 91, 92, 93, 94, 99, 100
sepsis, 85, 87, 89, 91, 92, 93, 94, 96, 99, 100
serum, 4, 6, 10, 11, 13, 16, 21, 28, 30, 34, 35, 48, 83, 88, 90, 91, 92, 93, 94, 95, 99
side effects, ix, 52, 53, 54, 67, 69, 85
signal transduction, 62, 89
signaling pathway, 5, 57, 63, 66, 69, 70, 71
sirtuin-1, 57, 71
sleep disorders, 16
sleep disturbance, 16, 39
sodium, ix, 9, 12, 15, 18, 35, 81
stress, 3, 6, 9, 16, 22, 33, 44, 59, 61, 66, 69, 72, 73, 74, 84, 89, 91, 92, 101
supplementation, ix, 4, 7, 8, 10, 13, 14, 16, 18, 19, 22, 23, 25, 34, 36, 38, 39, 44, 82, 86, 87, 88, 90, 91, 92, 93, 94, 97, 98, 99, 100
survival, viii, 44, 51, 53, 57, 92, 99

symptoms, 8, 10, 14, 15, 24, 25

T

therapy, 2, 19, 35, 37, 41, 59, 77, 83, 90, 97, 100
tissue, 5, 16, 58, 61, 89
toxicity, ix, 3, 15, 16, 25, 28, 52, 60, 67, 75, 99
trace element, viii, ix, 2, 3, 4, 5, 9, 12, 17, 18, 21, 23, 24, 25, 26, 27, 28, 30, 37, 38, 39, 43, 81, 82, 95
transplant, viii, 2, 3, 15, 19, 20, 21, 24, 25, 40, 41, 42, 43
transplant recipients, 19, 20, 21, 24, 40, 41, 42, 43
transplantation, 21, 28, 37, 42, 48, 65
treatment, 2, 8, 14, 15, 16, 18, 22, 24, 48, 53, 58, 59, 60, 62, 64, 65, 68, 72, 73, 76, 83, 85, 86, 88, 91, 93, 95, 96

U

upper respiratory tract, 86, 96

V

vitamin, vii, ix, 4, 5, 7, 9, 11, 12, 13, 14, 19, 20, 22, 23, 24, 25, 33, 35, 36, 37, 41, 42, 43, 44, 61, 82, 83, 84, 85, 86, 87, 88, 89, 90, 95, 96, 97, 98
vitamin A, 5, 9, 23, 25, 83, 90
vitamin B1, 5, 12, 13, 19, 24, 35, 36
vitamin B12, 13, 19, 24, 36
vitamin B12 deficiency, 19, 25
vitamin B6, 9, 13, 19, 20, 22, 23
vitamin B6 deficiency, 13, 20
vitamin C, ix, 5, 7, 9, 14, 20, 23, 24, 33, 37, 42, 82, 84, 85, 86, 87, 88, 90, 95, 96, 97
vitamin C deficiency, 7, 14, 25, 84
vitamin D, ix, 5, 82, 87, 88, 89, 97, 98
vitamin D deficiency, 87, 88, 89
vitamin E, 5, 19, 20, 22, 24, 82, 84, 89, 90, 98
vitamin K, 23
vitamin supplementation, 7, 25, 44

W

wound healing, 3

Z

zinc, ix, 3, 4, 5, 6, 7, 8, 9, 10, 11, 12, 14, 16, 17, 18, 21, 23, 24, 25, 27, 28, 29, 30, 33, 34, 35, 39, 40, 42, 57, 81, 82, 84, 91, 92, 93, 94, 98, 99, 100